INFINITY AND TRUTH

LECTURE NOTES SERIES
Institute for Mathematical Sciences, National University of Singapore

Series Editors: Chitat Chong and Wing Keung To
Institute for Mathematical Sciences
National University of Singapore

ISSN: 1793-0758

Published

Vol. 14   Computational Prospects of Infinity — Part I: Tutorials
edited by Chitat Chong, Qi Feng, Theodore A Slaman, W Hugh Woodin
& Yue Yang

Vol. 15   Computational Prospects of Infinity — Part II: Presented Talks
edited by Chitat Chong, Qi Feng, Theodore A Slaman, W Hugh Woodin
& Yue Yang

Vol. 16   Mathematical Understanding of Infectious Disease Dynamics
edited by Stefan Ma & Yingcun Xia

Vol. 17   Interface Problems and Methods in Biological and Physical Flows
edited by Boo Cheong Khoo, Zhilin Li & Ping Lin

Vol. 18   Random Matrix Theory and Its Applications:
Multivariate Statistics and Wireless Communications
edited by Zhidong Bai, Yang Chen & Ying-Chang Liang

Vol. 19   Braids: Introductory Lectures on Braids, Configurations and Their
Applications
edited by A Jon Berrick, Frederick R Cohen, Elizabeth Hanbury,
Yan-Loi Wong & Jie Wu

Vol. 20   Mathematical Horizons for Quantum Physics
edited by H Araki, B-G Englert, L-C Kwek & J Suzuki

Vol. 21   Environmental Hazards:
The Fluid Dynamics and Geophysics of Extreme Events
edited by H K Moffatt & E Shuckburgh

Vol. 22   Multiscale Modeling and Analysis for Materials Simulation
edited by Weizhu Bao & Qiang Du

Vol. 23   Geometry, Topology and Dynamics of Character Varieties
edited by W Goldman, C Series & S P Tan

Vol. 24   Complex Quantum Systems: Analysis of Large Coulomb Systems
edited by Heinz Siedentop

Vol. 25   Infinity and Truth
edited by Chitat Chong, Qi Feng, Theodore A Slaman & W Hugh Woodin

*For the complete list of titles in this series, please go to
http://www.worldscientific.com/series/LNIMSNUS

Lecture Notes Series, Institute for Mathematical Sciences,
National University of Singapore

Vol.
25

INFINITY AND TRUTH

Editors

Chitat Chong
Qi Feng
National University of Singapore, Singapore

Theodore A. Slaman
W. Hugh Woodin
University of California, Berkeley, USA

World Scientific

NEW JERSEY · LONDON · SINGAPORE · BEIJING · SHANGHAI · HONG KONG · TAIPEI · CHENNAI

Published by

World Scientific Publishing Co. Pte. Ltd.

5 Toh Tuck Link, Singapore 596224

USA office: 27 Warren Street, Suite 401-402, Hackensack, NJ 07601

UK office: 57 Shelton Street, Covent Garden, London WC2H 9HE

Library of Congress Cataloging-in-Publication Data

Infinity and truth / edited by Chitat Chong (National University of Singapore, Singapore), Qi Feng (National University of Singapore, Singapore), Theodore A. Slaman (University of California, Berkeley, USA), and W. Hugh Woodin (University of California, Berkeley, USA).

     pages cm. -- (Lecture notes series (Institute for Mathematical Sciences, National University of Singapore) ; volume 25)

  "Based on the talks given at the Workshop on Infinity and Truth held at the Institute for Mathematical Sciences, National University of Singapore, from 25 to 29 July 2011."

  Includes bibliographical references.

  ISBN 978-9814571036 (hardcover : alk. paper)

  1. Logic, Symbolic and mathematical--Congresses. 2. Mathematics--Philosophy--Congresses. 3. Set theory--Congresses. 4. Axiomatic set theory--Congresses. I. Chong, C.-T. (Chi-Tat), 1949– editor of compilation. II. Feng, Qi, 1955– editor of compilation. III. Slaman, T. A. (Theodore Allen), 1954– editor of compilation. IV. Woodin, W. H. (W. Hugh), editor of compilation.

  QA9.A1I54 2013

  510.1--dc23

2013041952

British Library Cataloguing-in-Publication Data

A catalogue record for this book is available from the British Library.

Printed in Singapore

CONTENTS

Foreword   vii

Preface   ix

Section I. Invited Lectures

Absoluteness, Truth, and Quotients   1
    Ilijas Farah

A Multiverse Perspective on the Axiom of Constructibility   25
    Joel David Hamkins

Hilbert, Bourbaki and the Scorning of Logic   47
    A. R. D. Mathias

Toward Objectivity in Mathematics   157
    Stephen G. Simpson

Sort Logic and Foundations of Mathematics   171
    Jouko Väänänen

Reasoning about Constructive Concepts   187
    Nik Weaver

Perfect Infinities and Finite Approximation   199
    Boris Zilber

Section II. Special Session

An Objective Justification for Actual Infinity?   225
    Stephen G. Simpson

Oracle Questions   229
    Theodore Slaman and W. Hugh Woodin

FOREWORD

The Institute for Mathematical Sciences (IMS) at the National University of Singapore was established on 1 July 2000. Its mission is to foster mathematical research, both fundamental and multidisciplinary, particularly research that links mathematics to other efforts of human endeavor, and to nurture the growth of mathematical talent and expertise in research scientists, as well as to serve as a platform for research interaction between scientists in Singapore and the international scientific community.

The Institute organizes thematic programs of longer duration and mathematical activities including workshops and public lectures. The program or workshop themes are selected from among areas at the forefront of current research in the mathematical sciences and their applications.

Each volume of the *IMS Lecture Notes Series* is a compendium of papers based on lectures or tutorials delivered at a program/workshop. It brings to the international research community original results or expository articles on a subject of current interest. These volumes also serve as a record of activities that took place at the IMS.

We hope that through the regular publication of these *Lecture Notes* the Institute will achieve, in part, its objective of reaching out to the community of scholars in the promotion of research in the mathematical sciences.

August 2013

Chitat Chong
Wing Keung To
Series Editors

PREFACE

A *Workshop on Infinity and Truth* was held at the Institute for Mathematical Sciences from 25 to 29 July 2011. The theme of the Workshop was on basic foundational questions such as: (i) What is the nature of mathematical truth and how does one resolve questions that are formally unsolvable within ZFC, an example being the Continuum Hypothesis? And (ii) Do the discoveries in mathematics provide evidence favoring one philosophical view, such as Platonism or Formalism, over others? Leading experts representing areas across the mathematical and philosophical logic participated in this Workshop. Altogether there were ten lectures delivered by Ilijas Farah, Joel Hamkins, Leo A. Harrington, Kai Hauser, Menachem Magidor, Stephen G. Simpson, Jouko Väänänen, Nik Weaver, W. Hugh Woodin, and Boris Zilber. The Workshop concluded with a special session on problems that would drive progress in foundational research ("Oracle Questions").

This volume consists of two sections: (I) written articles of talks presented at the Workshop received by the editors, and (II) a contribution by Stephen G. Simpson as well as a compendium of foundational questions discussed during the concluding session prepared by Theodore A. Slaman and W. Hugh Woodin. The volume also includes an article submitted by A. R. D. Mathias, who was invited but not able to attend the Workshop.

The meeting was supported by a grant from the John Templeton Foundation and formed part of the larger program *Computational Prospects of Infinity II* hosted and funded by the Institute for Mathematical Sciences. We wish to record here our gratitude for the support.

August 2013

Chitat Chong
National University of Singapore, Singapore

Qi Feng*
National University of Singapore, Singapore

Theodore A. Slaman
University of California at Berkeley, USA

W. Hugh Woodin
University of California at Berkeley, USA

Editors

---

*Current address: Chinese Academy of Sciences, China.

ABSOLUTENESS, TRUTH, AND QUOTIENTS

Ilijas Farah

Department of Mathematics and Statistics
York University
4700 Keele Street
North York, Ontario
Canada M3J 1P3
and
Matematicki Institut, Kneza Mihaila 34
Belgrade, Serbia
ifarah@mathstat.yorku.ca

The infinite in mathematics has two manifestations. Its occurrence in analysis has been satisfactorily formalized and demystified by the ϵ-δ method of Bolzano, Cauchy and Weierstrass. It is of course the 'set-theoretic infinite' that concerns me here. Once the existence of an infinite set is accepted, the axioms of set theory imply the existence of a transfinite hierarchy of larger and larger orders of infinity. I shall review some well-known facts about the influence of these axioms of infinity to the everyday mathematical practice and point out to some, as of yet not understood, phenomena at the level of the third-order arithmetic. Technical details from both set theory and operator algebras are kept at the bare minimum. In the Appendix, I include definitions of arithmetical and analytical hierarchies in order to make this paper more accessible to non-logicians. In this paper I am taking a position intermediate between pluralism and non-pluralism (as defined by P. Koellner in the entry on large cardinals and determinacy of the Stanford Encyclopaedia of Philosophy) with an eye for applications outside of set theory.

1. Finitism, 'Countablism' and a Little Bit Further

Let me recall von Neumann's definition of the cumulative hierarchy. We define sets V_α for ordinals α by recursion, so that $V_\emptyset = \emptyset$ and $V_{\alpha+1}$ is the power-set of V_α for every α. If δ is a limit ordinal, we let $V_\delta = \bigcup_{\alpha<\delta} V_\alpha$.

The Power-Set Axiom asserts the existence of $V_{\alpha+1}$, granted V_α exists. The existence of V_δ for a limit ordinal δ follows from the Replacement Axiom. We therefore have an increasing collection of sets, indexed by all ordinals, that provides framework for all of mathematics as we know it. All of number theory is formalized within V_ω. Every countable set, such as $\mathbb{Z}$, $\mathbb{Q}+i\mathbb{Q}$, or the free group with two generators, has an isomorphic copy inside V_ω (where ω is the least infinite ordinal). These sets, as well as all real numbers (defined via Dedekind cuts) belong to $V_{\omega+1}$. The set of real numbers therefore belongs to $V_{\omega+2}$. If A is a separable metric structure, such as ℓ_2, Tsirelson's Banach space, or Cuntz algebra $\mathcal{O}_2$, then it has a countable dense subset A_0 that can be identified (equipped with all of its metric, algebraic, and relational structure) with an element of $V_{\omega+1}$. Therefore, A itself, identified with the equivalence classes of Cauchy sequences in A_0, belongs to $V_{\omega+3}$. This also applies to objects that are separable in at least one natural metric, e.g., II$_1$ factors (ℓ_2 metric) or multipliers of separable C*-algebras (strict metric). Quotients, coronas, ultrapowers, or double duals of separable objects, as well as their automorphism groups, all belong to $V_{\omega+n}$ for a relatively small natural number n. Therefore $V_{\omega+\omega}$ already provides framework for most of non-set-theoretic mathematics.[a] Nevertheless, $V_{\omega+\omega}$ is not a model of ZFC since it fails the Replacement Axiom.

Accepting the existence of the empty set and the assertion that every set has the power-set has as a consequence the existence of V_n for all natural numbers n. However, the cardinality of V_6 is roughly $10^{19,738}$. Current estimates take number of fundamental particles in the observable universe to be less than 10^{85}. While these estimates are based on our current understanding of physics and are therefore subject to change, this shows that we have no concrete model of V_6. Can we nevertheless assert that V_6 exists? Can we claim that the power-set axiom is true in the physical world? For example, consider the set X of all electrons contained in this sheet of paper at this very moment. Does the power-set of X exist? (The problem may be in the comprehension axiom, or rather the question whether X is a set?)

The fact that most physical laws are only approximately true does not diminish their usefulness in concrete applications. (As von Neumann pointed out in [54], the truth is much too complicated to allow anything but approximations.) Regardless of whether the set of all reals $\mathbb{R}$ (or any other infinite set) exists or not, its formal acceptance provides us with remark-

---

[a]Because one can define a branch of mathematics to be 'set-theoretic' if it is concerned with objects that do not belong to $V_{\omega+\omega}$.

able mathematical tools. Accepting the transfinite hierarchy, together with some substantial large cardinal axioms, may be comparably illuminating. Present lack of arguments pro this view is fortunately counterbalanced by the complete absence of arguments against it.

2. Independence

By Gödel's Incompleteness Theorems, every consistent theory T that includes Peano Arithmetic and has a recursive set of axioms is incomplete. Moreover, the sentence constructed by Gödel is a Π_1^0 arithmetical sentence (see the Appendix). This means that it asserts that every natural number n has a certain property that can can be verified by computation. Such a sentence can be independent from PA only if it is true, and it can be falsified only by a nonstandard natural number. Therefore, no consistent recursively enumerable theory can capture the truth of all Π_1^0 arithmetic sentences. A straightforward recursive construction produces a complete and consistent extension of PA that is Π_2^0. However, a non-recursive axiomatization can hardly be considered satisfactory or natural.

A theory T is *1-consistent* if all Σ_1^0 sentences provable in T hold in the standard model of arithmetic $(\mathbb{N}, +, \cdot, 0)$. Every extension of ZFC that has an ω-model (i.e., a model whose set of natural numbers is standard) is 1-consistent. In particular, all 'reasonable' theories obtained by adjoining large cardinal axioms to ZFC are 1-consistent. Moreover, the relative strength of large cardinal axioms is accurately measured by the inclusion between sets of their Σ_1^0 consequences. This is because if ϕ and ψ are large cardinal axioms and ϕ is stronger than ψ then ϕ implies that $\neg\psi$ is not consistent. Via the Gödel coding, a statement $\mathrm{Con}(\neg\psi)$ is equivalent to a Π_1^0 statement and by Gödel's Incompleteness theorem it does not follow from ψ.

Open problems of number theory could turn out to be independent from ZFC. Or rather, the only known way to prove that this is not the case seems to be to either prove or refute each one of them. However, the standard set theoretic tools for proving independence cannot prove independence of arithmetical statements (a notable exception here is the line of research starting with Paris–Harrington theorem developed further by Harvey Friedman [28]). This also applies to mathematical statements that have equivalent reformulation that is an arithmetical statement, such as for example the Kadison–Singer problem.

The existence of true arithmetic statements that are not provable in any recursive extension of ZFC can be considered as a deficiency of the first-order logic. However, the generally accepted notion of 'proof' in mathematics is the Hilbert-style proof in first-order logic, from some standard axiomatization of mathematics such as ZFC. Other prominent competing formalizations of mathematics, such as the intuitionism, are weaker than the first-order logic. The first-order logic is also model-theoretically the best-behaved of all logics, as Lindström's theorem ([9]) characterizes the first-order logic as the strongest logic that satisfies both compactness and Löwenheim–Skolem theorems.

Since all well-founded models of PA are isomorphic, assuming that $\mathbb{N}$ exists (i.e., the axiom of infinity) implies that each arithmetical statement has a concrete truth value, regardless of whether it is independent from PA. The situation with the statements of the second-order arithmetic, i.e., those that require quantification over sets of natural numbers, is more complicated. Sets of natural numbers are naturally identified with elements of the Cantor set, or real numbers, or elements of any recursively defined Polish space. This extends to objects that can be coded by sets of natural numbers, such as Borel sets or separable C*-algebras. Therefore most mathematical statements about separable objects that are elements of a standard Borel space are really statements of the second-order arithmetic.

At this point I will digress to retell the well-known story of one of the major achievements of pure set theory to this date.

2.1. *The story of projective sets*

One does not know, and one will never know, of the family of projective sets, although it has cardinality $2^{\aleph_0}$ and consists of effective sets, whether every member has cardinality $2^{\aleph_0}$ if uncountable, has the Baire property, or is even Lebesgue measurable.

N. Luzin, 1925

Descriptive set theory started with the work of Lebesgue, who initiated the systematic study of definable sets of real numbers (see the introduction to [40]). Classical regularity properties of sets of reals, such as Lebesgue measurability, property of Baire, and the perfect set property, were quickly established for Borel, or equivalently $\mathbf{\Delta}_1^1$, sets. Famously, Lebesgue claimed that continuous images of Borel sets are Borel. This assertion implies that all sets of reals that are first-order definable in second-order arithmetic have all the classical regularity properties. It also implies that $0 = 1$ but the proof is slightly longer. The importance of, and benefit from, formal mistakes of this sort in mathematics can hardly be overstated.[b]

Fortunately, the truth happens to be much more exciting. Suslin noticed that Lebesgue's assertion was false and named continuous images of Borel sets analytic sets. Suslin's result gave rise to a proper hierarchy of projective sets. Lebesgue measurability and property of Baire of analytic sets was quickly established, but the progress was stalled at the very next level, of continuous images of complements of analytic sets (i.e., Σ_2^1 sets).

In 1925, when Luzin articulated his pessimistic thought, the central question of Descriptive Set Theory was whether Σ_2^1 sets are Lebesgue measurable. Fifteen years later, a partial solution to the problem was found. Kurt Gödel introduced his constructible universe, L, and proved that it has a Σ_2^1 well-ordering of the reals. This well-ordering also has the property that all of its proper initial segments are countable. A Fubini argument immediately implies that the well-ordering itself (when considered as a subset of the square) is not measurable. In addition, any transfinite construction of a set of reals that uses Continuum Hypothesis could be ran along this Σ_2^1 well-ordering to produce projective sets, resulting in projective sets with even more peculiar properties.

This only demonstrates that one cannot prove that projective sets have classical regularity properties starting from ZFC. Gödel himself never considered this as a solution to the problem of regularity properties of projective sets of reals.

Another quarter of a century later, after the advent of forcing, Solovay constructed a model of ZFC in which all projective sets have all classical regularity properties. Unlike Gödel's, Solovay's theorem still allows a theoretical possibility that one can construct a non-measurable projective set of reals starting from ZFC alone. This is because Solovay's model was constructed from a model of ZFC in which there exists an inaccessible cardinal, and the existence of such a model is a strictly stronger assumption than the existence of a model of ZFC, used by Gödel. By a result of Shelah, the use of an inaccessible cardinal is necessary in Solovay's construction: Starting from a model in which all Σ_3^1 sets are measurable one can construct a model of ZFC with an inaccessible cardinal. This, in particular, implies that in a strictly formal sense this result is weaker than Gödel's. While by the latter result in ZFC one cannot prove that all projective sets are Lebesgue measurable, the possibility that one can construct a non-measurable set of

---

[b]An entertaining account can be found in [3].

reals in ZFC cannot be ruled out by a formal reasoning starting from ZFC.

In light of these results it is fair to say that Luzin had little chance of resolving the question of regularity properties of projective sets, as such a resolution necessitates intricate metamathematical considerations unforeseen in Luzin's time. However, the truth turns out to be even more compelling.

By a 1988 result of Martin and Steel, the existence of infinitely many Woodin cardinals implies Projective Determinacy. On the other hand, Woodin proved that Projective Determinacy implies the existence of models with arbitrarily large (finite) number of Woodin cardinals.

Therefore the question whether projective sets are Lebegue measurable has been completely answered from the formal point of view. The axioms of ZFC are too weak to provide an answer.[c] Nevertheless, the picture provided by large cardinals is so coherent, robust, and beautiful that Projective Determinacy has been accepted as a bona fide axiom, at least by the California school of set theory (see [57]).

But let us now go back to 1920's and Luzin. He observed that proving measurability of Σ_1^1 or Π_1^1 sets was not substantially more difficult than proving measurability of Borel sets. An attempt at proving that continuous images of these sets are measurable resulted in unsurmountable technical difficulties, a complete resolution of which required development of iteration trees for inner models of Woodin cardinals, which is some of the deepest and most intricate set theory as we know it. Therefore Luzin's '...and one will never know...' was an educated, and even lucid (albeit somewhat arrogant) assessment of the problem.

3. Absoluteness

Absoluteness phenomena are calibrated by the Levy hierarchy of formulas ([38]). Levy proved that Σ_1 statements are absolute between transitive models of ZFC. This was followed by Shoenfield who proved the absoluteness of Σ_2^1 statements between models of ZFC containing all countable ordinals (see the Appendix). This result predates results discussed in §2.1 by two decades. By Shoenfield's result, any statement in the language of ZFC that is provably equivalent to a Σ_2^1 statement is absolute between 'reasonable' models of ZFC (reasonable meaning well-founded and containing all countable ordinals) and therefore arguably has a concrete truth value.

---

[c]Unless it turns out that the existence of infinitely many Woodin cardinals leads to a contradiction, a possibility that cannot be formally ruled out.

I now proceed to give some flavour of the technical aspects of mathematics involved in this. A *tree* on a set X is a set of finite sequences of elements of X closed under taking initial segments. It is ordered by the end-extension. The use of trees in the following argument is the basis for practically all presently known absoluteness results.

Lemma 3.1: *Let M be a transitive model of a large enough fragment of ZFC. If T is a tree in M, then T has an infinite branch in M if and only if it has an infinite branch.*

Proof: In M recursively define a rank function from T into the ordinals by setting $\rho(s) = 0$ if s is a terminal node of T, and

$$\rho(t) = \sup(\rho(s) + 1)$$

where s ranges over all immediate successors of t, i.e., nodes of the form $t^\frown x$ for some $x \in X$. This defines a strictly decreasing function from a subset of T into the ordinals by transfinite recursion. Moreover, the complement of $\mathrm{dom}(\rho)$ has no terminal nodes. Therefore, if it is nonempty then by the axiom of Dependent Choices (a rather weak consequence of the Axiom of Choice) T has an infinite branch.

Therefore, if T has no infinite branches then there is a strictly decreasing function from T into the ordinals, and such a function prevents T from acquiring an infinite branch in any larger (well-founded) universe. $\square$

A tree with no branches is said to be *well-founded* (visualize the tree growing downwards). Hence Lemma 3.1 states that the well-foundedness of trees is absolute between transitive models of a large enough fragment of ZFC.

3.1. *Beyond projective sets*

A closer introspection of the proof of Lemma 3.1 shows that the salient feature of analytic sets A is the fact that there are trees T and S such that (i) T projects to A, (ii) projection of S is disjoint from the projection of T, and (iii) every real number is in the projection of one of these trees. The relevant tree S is a tree of attempts to construct rank function on T (see [32]). A subset A of a Polish space is *universally Baire* ([26], [36]) if there exist trees T and S satisfying (i)–(iii) in all set forcing extensions (such trees are necessarily proper classes).

As pointed out by Steel ([25]), the invariance of the theory of $L(\mathbb{R})$ under set forcing (an assertion that implies PD) is equivalent to the existence of an iterable model with ω Woodin cardinals $\mathcal{M}_\omega$. The latter is a countable transitive structure, and therefore coded by a real number. The iterability of $\mathcal{M}_\omega$ is witnessed by a set of real numbers that is universally Baire ([26]). Therefore all transitive models of ZFC (and even all ω-models of ZFC) containing $\mathcal{M}_\omega$ that are correct about its iteration strategy (in a certain technical sense, see [58]) agree about the theory of projective sets.

For ZFC, or some stronger theory, one can consider only (appropriately defined) reasonable models of the theory and accept statements that hold in all reasonable models as true. This is Woodin's approach to Ω-logic ([58]), in which he defines a class of 'test structures' and writes $T \models_\Omega \phi$ if ϕ holds in all test structures in which T holds. The technical definition of Woodin's Ω-logic involves being A-closed for a universaly Baire set A, and it is preserved by set-forcing. In case of ZFC, the minimum requirement on a test structure M is that it is a transitive model of ZFC. This is captured by the notion of β-*logic*. One defines $T \models_\beta \phi$ if every transitive model of theory T satisfies ϕ. By Lemma 3.1, β-logic decides all Σ_1^1 statements, but unfortunately there is no known notion of a proof in β-logic.

A 'phase transition' for models of ZFC occurs when the following large cardinal axiom is assumed. I will not define Woodin cardinals; a non-technical (to the extent it is possible) introduction is given in [50].

($\mathbb{W}_\infty$) There exist arbitrarily large Woodin cardinals.

This assumption is robust under forcing and implies the existence and iterability of $\mathcal{M}_\omega$. Therefore $\mathbb{W}_\infty$ implies all the regularity properties of projective sets. It also implies that the σ-algebra of universally Baire sets ([26]) is closed under projections and that all of these sets share all the regularity properties, including one of the strongest determinacy axioms presently known, $AD_\mathbb{R} + \Theta$ is regular (see [45] for a discussion of Solovay hierarchy).

Absoluteness results for universally Baire sets can be rephrased by expanding the language. Let $\mathcal{L}_{\mathrm{uB}}$ denote the language obtained from the language of ZFC by adding predicates for all universally Baire sets and (redundantly) constants for all real numbers. Then $\mathbb{W}_\infty$ implies that all formulas of $\mathcal{L}_{\mathrm{uB}}$ are absolute for set-forcing extensions. It also implies that all projective formulas of $\mathcal{L}_{\mathrm{uB}}$ (i.e., all Σ_n^1 formulas using predicates for universally Baire sets) are also absolute for set-forcing extensions. However, $\mathbb{W}_\infty$ implies that the σ-algebra of universally Baire sets is closed under continuous images and therefore this is not a genuine strengthening (see [36]).

3.2. *Absoluteness and the uncountable*

There is a notable absoluteness result which goes both beyond the second-order arithmetic and the theory of universally Baire sets and which does not require any large cardinal assumptions. It is a corollary of Keisler's completeness theorem for non-standard logic $L_{\omega_1\omega}(Q)$ ([33]). This is the extension of first-order logic obtained by allowing countable conjunctions and disjunctions, as well as the additional quantifier Qx which is interpreted as 'there exist uncountably many x.' Keisler introduced a finite set of logical axioms for $L_{\omega_1\omega}(Q)$ and proved that an $L_{\omega_1\omega}(Q)$-sentence ϕ has a standard model (i.e., a model in which Qx has its intended interpretation) if and only if its negation $\neg\phi$ is not provable. While this logic has an infinite rule of inference (handling infinite conjunctions), proofs are well-founded trees tagged with formulas. Therefore by Lemma 3.1 the existence of uncountable structures whose properties can be described in $L_{\omega_1\omega}(Q)$ is absolute between transitive models of a large enough fragment of ZFC. A typical consequence states that for any fixed Borel subset B of $\mathbb{R}^n$ the assertion that there exists an uncountable set X such that $X^n \subseteq B$ is absolute between transitive models of ZFC. Every Borel set can be coded by an $L_{\omega_1\omega}$ sentence that describes its recursive definition, using some fixed recursive basis of open sets. The case $n = 2$ was used by Shelah to prove that the countable chain condition is absolute for Borel forcing notions.

Let us now expand $\mathcal{L}_{uB}$ by adding a predicate for the nonstationary ideal on ω_1 (in this context we add this predicate primarily because we can). We shall denote this language by $\mathcal{L}_{uB,NS}$. Via the work of Woodin ([58]), $\mathbb{W}_\infty$ implies a strengthening of Keisler's absoluteness theorem. If ϕ is a sentence $\mathcal{L}_{uB,NS}$ such that both ϕ and $\neg\phi$ are provably equivalent to a Σ_2 sentence, then the assertion $H(\aleph_2) \models \phi$ cannot be changed by set forcing (see [21], where other absoluteness results along similar lines were considered) (see the Appendix for definition of $H(\kappa)$). Therefore $\mathbb{W}_\infty$ implies that the provably Δ_2-fragment of the $\mathcal{L}_{uB,NS}$ theory of $H(\aleph_2)$ is absolute under set forcing. This implies, for example, that for any projective, or even universally Baire $B \subseteq \mathbb{R}^{<\mathbb{N}}$ (the finite sequences of reals) the existence of an uncountable X such that $X^{<\mathbb{N}} \subseteq B$ is absolute under set forcing (see [21] for other applications).

3.3. *Level by level*

The relative consistency of a sentence ϕ is, via Gödel coding, equivalent to a Π_1^0 arithmetic statement. Therefore an oracle with access to Π_1^0 theory of the natural numbers would give full insight into the relative consistency of large cardinal axioms. However, the relevant question (from the point of view of Descriptive Set Theory) is that of the existence of (necessarily well-founded) iterable models of large cardinal axioms. The existence of well-founded models is a statement of the second-order arithmetic. However, the iterability of these models is witnessed by (ideally universally Baire) sets of reals. The existence of such sets is already the statement of the third-order arithmetic. However, their universal Baireness is witnessed by class-sized trees, and therefore it transcends any fixed V_α.

4. Third-Order Arithmetic

We move one level up and consider the third-order arithmetic. With the identification of sets of natural numbers with the reals, this means that we are quantifying over sets of real numbers. More generally, if X is any Polish space ([32]) then we allow quantification over subsets of X.

The setting for the third-order arithmetic is provided by the structure $H(\mathfrak{c}^+)$ of all sets whose transitive closure has cardinality no larger than $\mathfrak{c}$, the cardinality of $\mathbb{R}$. All Polish spaces and all of their subsets have a isomorphic copies inside this structure. A compelling motivation for study of $H(\mathfrak{c}^+)$ is given by the fact that this is a reasonably small structure in which most of the mainstream mathematical practice takes place.

4.1. *Conditional absoluteness*

A statement is Σ_1^2 if it is of the form

$$(\exists X \subseteq \mathbb{R})\phi(X)$$

where ϕ is an analytical statement (see Appendix §A.3) of the language $\mathcal{L}_{\mathrm{uB,NS}}$. An example of a Σ_1^2 statement is the Continuum Hypothesis, CH, since it is equivalent to the assertion that there exists a well-ordering of $\mathbb{R}$ all of whose proper initial segments are countable.

Every model of ZFC has a set-forcing extension in which CH fails and a set-forcing extension in which CH holds. Therefore the unconditional absoluteness results of §3.2 cannot be extended to Σ_1^2 statements.

A large cardinal axiom substantially stronger than $\mathbb{W}_\infty$ implies that a new phenomenon takes place.

($\mathbb{MW}_\infty$) There exist arbitrarily large measurable Woodin cardinals.

Although the existence of a Woodin cardinal implies the existence of many measurable cardinals, a Woodin cardinal is not necessarily measurable. While $\mathbb{W}_\infty$ is a relatively mild large cardinal assumption with well-understood inner models by results of Mitchell-Steel and Neeman ([41]), the assumption that there exists a single measurable Woodin cardinal is far beyond the reach of the present inner model theory. Woodin proved ([56], for a proof see [36]) that $\mathbb{MW}_\infty$ implies all statements of the form CH $\Rightarrow \phi$, for a Σ_1^2-statement ϕ in $\mathcal{L}_{uB,NS}$, are set-forcing absolute. Even stronger, for every Σ_1^2 sentence ϕ in $\mathcal{L}_{uB,NS}$ we have either that $\models_\Omega$ CH $\Rightarrow \phi$ or $\models_\Omega \neg\phi$.

One should note that this result is not stating that a Σ_1^2 statement is either inconsistent or it follows from CH. It is not even stating that $\mathbb{MW}_\infty$ implies every Σ_1^2 statement is either inconsistent or it follows from CH; there are easy metamathematical counterexamples to this assertion. What it says is that $\mathbb{MW}_\infty$ implies that if consistency of a Σ_1^2 statement can be proved in ZFC (possibly augmented by a large cardinal axiom) by standard presently available set-theoretic tools (i.e., by forcing—but not by passing to an inner model such as Gödel's L), then it follows from CH. Note that this excludes proofs obtained by forcing over L, or a model of some other anti-large cardinal axiom.

Remarkably, this theorem has practical consequences. For example, its consequence is that $\mathbb{MW}_\infty +$ CH imply for any Σ_1^2 formula $\phi(x)$ in $\mathcal{L}_{uB,NS}$ the set $\{x \in \mathbb{R} : \phi(x)\}$ is Lebesgue measurable and has the property of Baire. This is proved by using Solovay's amoeba forcing argument ([49]).

It was conjectured by John Steel that a sufficiently strong large cardinal assumption implies analogous result for $\Diamond$ in place of CH and for all Σ_2^2 sentences ϕ: either $\models_\Omega \Diamond \Rightarrow \phi$ or $\models_\Omega$ CH $\Rightarrow \neg\phi$ (note that the more simple-minded analogy is outright false, since $\neg$ CH is Π_1^2 and therefore Σ_2^2. This conjecture is wide open and the current understanding of the Inner Model Theory suggests that its confirmation may require very substantial large cardinal axioms ([59]). A result of Magidor and Malitz ([39]) is relevant to this problem. Logic $L_{\omega_1\omega}(Q^{<\omega})$ can be extended by adding Ramseyan quantifiers $(Q^n x_1, \ldots, x_n)$ for all $n \in \mathbb{N}$. Formula $(Q^n x_1, \ldots, x_n)\phi(x_1, \ldots, x_n)$ is interpreted as 'there exists an uncountable set X such that $\phi(x_1, \ldots, x_n)$ holds for every n-tuple of elements $x_1, \ldots, x_n$ in X.' In [39] a completeness theorem for this logic analogous to Keisler's was proved, assuming $\Diamond$. This shows that for a restricted class of Σ_2^2 sentences Steel's question has a positive answer. An extension of Magidor–Malitz result for $\mathcal{L}_{uB,NS}$ which also allows Ramseyan quantifier of unbounded dimension was proved from $\mathbb{MW}_\infty$ in [20]. An another test question for Steel's conjecture was answered affirmatively by Woodin: $\mathbb{MW}_\infty$ implies there exists a model of ZFC in which all Σ^2_2 statements that can be forced to hold with $\Diamond$ are simultaneously true ([34]).

4.2. Π^2_1?

Is there a conditional absoluteness result dual to Woodin's Σ^2_1 absoluteness theorem, based on some other axiom instead of the Continuum Hypothesis? We need to put this question in a proper context. One has to note that in the presence of CH Σ^2_1 statements are really Σ_1 statements of $H(\aleph_2)$. The proof of Woodin's theorem hinges of this fact—it proceeds by starting from a countable model M of the desired Σ^2_1 statement and constructing an elementary chain of length ω_1 of well-founded models extending M. Strong large cardinal assumption is needed to assure that every real can be absorbed in the direct limit of this chain.

Definition 4.1: If $\mathbb{H}$ is a given class of Hausdorff spaces, then $\mathrm{FA}(\mathbb{H})$ is the assertion that for every $\Omega \in \mathbb{H}$ the intersection of any collection of $\aleph_1$ dense open subsets of Ω is dense.

Even the weakest nontrivial forcing axiom, $\mathrm{FA}(\{[0,1]\})$, clearly contradicts CH. (This axiom is true in Cohen's original model for the negation of the Continuum Hypothesis.) The popular Martin's Axiom. $\mathrm{MA}_{\aleph_1}$, is the prototype for forcing axioms. Martin's Maximum, MM, ([27]) is provably the strongest forcing axiom consistent relative to large cardinals. MM^{++} is a technical strengthening of MM.

In [58] Woodin proved that $\mathbb{W}_\infty$ implies there exists a model that maximizes the Π_2-theory of $H(\aleph_2)$ in $\mathcal{L}_{\mathrm{uB,NS}}$. This model is obtained by forcing over an inner model of the Axiom of Determinacy. A salient feature of this model is that it an extension of a model $L(\Gamma, \mathbb{R})$ of the Axiom of Determinacy by a homogeneous forcing notion. By the homogeneity of this forcing, the theory of the model is decided in the ground model $L(\Gamma, \mathbb{R})$. But $\mathbb{W}_\infty$ implies that the theory of $L(\Gamma, \mathbb{R})$ is not changeable by forcing, and therefore the theory of Woodin's model is canonical. Woodin's axiom $(*)$ captures the essence of this model.

This axiom and forcing axioms have very similar effect to the truth of Π^2_1 statements. As a matter of fact, a bounded version of the latter, $\mathrm{MM}^{++}(\mathfrak{c})$ (a Π^2_1 statement in language with a predicate for the nonstationary ideal on ω_1), is a consequence of a strong version of $(*)$ ([58]). An important open problem is to reconcile the theories of (*) and forcing axioms (see [37]). At present it is not known whether the conjunction of these axioms is consistent. The only available models of (*) are obtained only by forcing over models of the Axiom of Determinacy, AD, (where the Axiom of Choice necessarily fails) or over some fine-structural models of relatively mild large cardinal axioms by using the extender algebra (see [19]). Presently known models of AD fall short of having the consistency strength of supercompact cardinal, that appears to be needed for forcing MM^{++} (see [53]). However, in [45] Sargsyan conjectured the existence of models of determinacy that capture arbitrarily large cardinal axioms. A confirmation of Sargsyan's conjecture could be a step towards proving the joint consistency of (*) and strong forcing axioms.

Problem 4.2: *Find a reasonably strong analogue of Woodin's Σ_1^2 absoluteness theorem for some class of Π_1^2 statements based on MM^{++} or (*).*

Large cardinal strength of forcing axioms largely stems from their reflection properties. Bounded forcing axioms are technical weakenings of forcing axioms with the reflection component diminished (but still present to some extent). The precise definition of bounded forcing axioms can be omitted here since in [4], Bagaria proved that they can be reformulated as principles of generic absoluteness. By his results, a bounded forcing axiom for class $\mathbb{H}$ is equivalent to the assertion that for every stationary preserving forcing notion $\mathbb{P}$ in class $\mathbb{H}$ every $a \in H(\aleph_2)$, and every Σ_1 formula $\phi(x)$ of $\mathcal{L}_{uB,NS}$ the truth of the statement $H(\aleph_2) \models \phi(a)$ cannot be changed by $\mathbb{P}$. In symbols,

$$H(\aleph_2)^V \prec_1 H(\aleph_2)^{V^{\mathbb{P}}}.$$

See also [11]. As pointed out earlier, by Woodin's results $\mathbb{W}_\infty$ implies that the Σ_2-theory of $H(\aleph_2)$ expressed in $\mathcal{L}_{uB,NS}$ cannot be changed by forcing.

A result along similar lines was proved by Viale ([52]) from a very strong large cardinal assumption. If there are class many supercompact cardinals and MM^{++} holds and $\mathbb{P}$ is stationary-preserving forcing that forces MM^{++} then $H(\aleph_2)^V \prec H(\aleph_2)^{V^{\mathbb{P}}}$ in $\mathcal{L}_{uB,NS}$. In other words, MM^{++} fixes the theory of $H(\aleph_2)$ in a way similar to the way that $\mathbb{W}_\infty$ fixes the theory of $L(\mathbb{R})$. The resolution of Problem 4.2, if any, will be of a rather restricted nature. A number of Π_1^2 statements relatively consistent with ZFC and large cardinals imply that MM, or even the weaker $MA_{\aleph_1}$, fails. Among such statements are $\mathfrak{b} < \mathfrak{d}$ or 'there are no P-points in $\beta\mathbb{N}$.' Here is a working conjecture.

Conjecture 4.3: *Assume* MW_∞. *If* ϕ *is* $\mathcal{L}_{\mathrm{uB,NS}}$ *sentence which is both* Π_2 *and* Π_1^2 *and it is true in some forcing extension, then* ϕ *is true in every forcing extension that satisfies* MM^{++}.

A rather coherent rigidity theory of quotient structures (see §5) developed over the last thirty years suggests that such conditional absoluteness theorem is true for some restricted class of Π_1^2 sentences. I will proceed to describe a set of results that appear to be instances of a yet unknown general rigidity theorem.

5. Quotient Borel Structures

I will now make a deliberately vague definition that encompasses some phenomena observed in areas of mathematics fairly distant from one another (or so we thought).

5.1. *Trivial automorphisms*

Assume $\mathfrak{A}$ and $\mathfrak{B}$ are models of the same signature whose universes are Polish spaces and all relations and functions are continuous functions. Further assume that I and J are Borel ideals of $\mathfrak{A}$ and $\mathfrak{B}$, respectively and

$$\Phi\colon \mathfrak{A}/I \to \mathfrak{B}/J$$

is an isomorphism. We say Φ is *topologically trivial* if there exists a Borel-measurable map $\Psi\colon \mathfrak{A} \to \mathfrak{B}$ such that the diagram commutes.

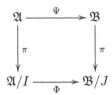

This is a bit weaker than the triviality requirement given in [12] and [16], where it was required that in addition Ψ is a homomorphism. The question when 'Borel' triviality implies this 'algebraic' triviality was studied in the context of both groups and Boolean algebras ([30], [31]). As interesting as they are, these 'Ulam stability' problems have little to do with the theme pursued in this article, in particular because the assertion 'every topologically trivial homomorphism between $\mathfrak{A}/I$ and $\mathfrak{B}/J$ is trivial' is Π_2^1 and therefore absolute (see the Appendix). My reason for considering only the Borel triviality in the present context is that in the case of C*-algebras it is still unclear what the right definition of algebraic triviality is (see the discussion in [10]).

These definitions provide a general framework for rigidity theory of Čech–Stone remainders, rigidity of analytic quotients and rigidity of coronas of C*-algebras (see [12], [16], [18] and below).

The assertion that two quotient Borel structures are isomorphic, or that a quotient Borel structure has a nontrivial automorphism, is a Σ_1^2 statement. On the other hand, the assertion that two quotient Borel structures are isomorphic via a trivial, or a topologically trivial, automorphism, is Π_2^1 in the codes for the structures.

5.2. *C*-algebras and their multipliers*

A particularly interesting and well-studied class of quotient structures are corona algebras of separable C*-algebras, also called *outer multiplier algebras* (see [6] for this and other background on operator algebras). Recall that an abstract C*-algebra is a complex Banach algebra A with involution * that satisfies the C*-equality, $\|a\|^2 = \|aa^*\|$ for all $a \in A$. A concrete C*-algebra is a norm-closed, self-adjoint subalgebra of the algebra $\mathcal{B}(H)$ of bounded operators on a complex Hilbert space H. Every concrete C*-algebra is an abstract C*-algebra and by the Gelfand–Naimark–Segal theorem every abstract C*-algebra is isomorphic to a concrete C*-algebra. The category of abelian C*-algebras is equivalent to the category of locally compact Hausdorff spaces. More precisely, every abelian C*-algebra A is isomorphic to $C_0(X)$, where X is the *spectrum* of A: the space of all *-homomorphisms of A into $\mathbb{C}$ (such homomorphisms are automatically continuous!) equipped with the weak* topology. If A is in addition unital, then X is compact and $A \cong C(X)$. *-homomorphisms between abelian C*-algebras contravariantly correspond to continuous maps between their spectra.

The operation of taking Čech–Stone compactification of a locally compact Hausdorff space has the construction of a multiplier algebra as its (non-functorial) analogue in the category of C*-algebras. If A is a subalgebra of $\mathcal{B}(H)$, then the *idealizer* of A is

$$\{b \in \mathcal{B}(H) : bA \subseteq A \text{ and } Ab \subseteq A\}.$$

This is easily checked to be a norm-closed, selfadjoint subalgebra of $\mathcal{B}(H)$. If moreover A acts nondegenerately on H (i.e., $bA = \{0\}$ only if $b = 0$), then the isomorphism type of this idealizer depends only on A, and not on the representation of A in $\mathcal{B}(H)$. Therefore one can define the *multiplier algebra* $M(A)$ of an abstract C*-algebra A to be the idealizer of some (any) of its nondegenerate representations. (By the Gelfand–Naimark–Segal theorem one has plenty of such representations.)

It is clear that $M(A) = A$ if A is unital. Also, $M(C_0(X))$ is isomorphic to $C_b(C_0(X))$, the algebra of bounded continuous functions on X, and the latter is in turn isomorphic to $C(\beta X)$, the space of continuous functions on the Čech–Stone compactification of X. By the construction, A is always a norm-closed, two-sided and self-adjoint ideal and the quotient algebra $Q(A) := M(A)/A$ is the *corona algebra* of A. Note that $M(C_0(X))/C_0(X)$ is isomorphic to $C(\beta X \setminus X)$, the space of continuous functions on the remainder (also called corona) of locally compact space X.

While the multiplier of a non-unital separable algebra is necessarily nonseparable, it does carry a natural Polish topology. The *strict* topology on $M(A)$ is the topology induced by the seminorms $b \mapsto \|baa^*\|$, for $a \in A$. If A is separable then this topology is induced by a single (carefully chosen) $a \in A$, and $d(b,c) = \|(b-c)a\|$ is a complete separable metric on $M(A)$ that has A as a dense subset. Therefore the corona $M(A)/A$ is subject to the considerations given in §5.1. For more on connections between C*-algebras and set theory see [55].

5.3. *General rigidity conjectures*

The following two conjectures are deliberately vague. As in §5, $\mathfrak{A}$ and $\mathfrak{B}$ are Polish models and I and J are their Borel ideals.

Conjecture 5.1: *If the assertion 'every isomorphism between $\mathfrak{A}/I$ and $\mathfrak{B}/J$ is trivial' can be forced then it follows from Martin's Maximum.*

Conjecture 5.2: *For a large class of Borel quotients $\mathfrak{A}/I$ and $\mathfrak{B}/J$, the assertion 'every isomorphism between $\mathfrak{A}/I$ and $\mathfrak{B}/J$ is trivial' is relatively consistent with ZFC.*

MM (and much less than MM) implies all isomorphisms between the following are trivial, and even algebraically trivial (see [46], [16], [51] for (1)–(2), [22] for (3), [14] for (4), and [18] and [17] for (5)).

(1) Boolean algebras $\mathcal{P}(\mathbb{N})/I$, for a large class of Borel ideals I.
(2) Boolean algebras $\mathcal{P}(\kappa)/\mathrm{Fin}$.
(3) Corona algebras $M(A)/A$, for A separable abelian C*-algebra generated by projections.

(4) Tensor products of algebras as in (3).

(5) Calkin algebra associated with an arbitrary Hilbert space.

Note that (2) and the nonseparable case of (5) belong to an extension of the context given in §5.1 to nonseparable yet 'definable' quotient structures. What assumptions on $\mathfrak{A}, \mathfrak{B}, I$ and J agree with the conjecture 'all isomorphisms between $\mathfrak{A}/I$ and $\mathfrak{B}/J$ are trivial?' In the known cases it is the existence of a (definable) partial ordering with rich gap structure on these quotients. An ability to freeze these gaps (i.e., to make them indestructible by $\aleph_1$-preserving forcing) appears as a requirement for Conjecture 5.2.

Here is an example (taken from [13]) that shows that in the category of groups nontrivial automorphisms can be constructed in ZFC. Consider the group $G = (\mathbb{Z}/2\mathbb{Z})^{\mathbb{N}}$ and let $G_0 = \bigoplus_{\mathbb{N}}(\mathbb{Z}/2\mathbb{Z})$. Then G/G_0 is a vector space over F_2 of dimension $2^{\aleph_0}$ and it therefore has $2^{\mathfrak{c}}$ automorphisms, most of them nontrivial. However, all automorphisms with a Borel-measurable representation are trivial ([13]).

A question closely related to the above is whether two Borel quotient structures are isomorphic, and what does the automorphism group of a Borel quotient structure look like? Special cases of this rigidity problem were asked in different categories. The case of quotient Boolean algebras $\mathcal{P}(\mathbb{N})/I$ was considered in [12] and there was not much progress on this question since [16]. The most general case of a question of this sort that is reasonably manageable by current methods is its C*-algebraic case. This is because by Gelfand–Naimark duality the categories of abelian C*-algebras and locally compact Hausdorff spaces are equivalent, and by the Stone duality the category of Stonian spaces is equivalent to the category of Boolean algebras.

These algebras belong to our framework of Borel quotient structures. Little is known about the question of P. W. Ng, when the isomorphism of coronas of separable C*-algebras A and B implies A and B are isomorphic? At present it is not known whether there are coronas that are isomorphic via a nontrivial isomorphism n ZFC. While nontrivial automorphisms of coronas have been constructed using the Continuum Hypothesis ([44], [10]), they are all 'locally trivial' (cf. Brown–Douglas–Fillmore problem on K-theory reversing automorphisms mentioned in §A.4.).

Under what assumptions is every automorphism of $\mathfrak{A}/I$ topologically trivial? In [46], Shelah constructed a model of ZFC in which all automorphisms of $\mathcal{P}(\mathbb{N})/\mathrm{Fin}$ are trivial. The feature of $\mathcal{P}(\mathbb{N})/\mathrm{Fin}$ most important for the proof is the existence of gaps. Every nontrivial automorphism Φ

is destroyed by adding a new element of $\mathcal{P}(\mathbb{N})/I$ whose image would necessarily fill a gap in $\mathcal{P}(\mathbb{N})/\text{Fin}$. Shelah's proof used *oracle-cc*, a delicate iteration of mild forcing notions, to assure that these gaps are not filled by subsequent forcings. The technique of *freezing gaps* ([47], [51]) is then used to conclude that forcing axioms imply the triviality of all automorphisms. This technique produces an object of cardinality $\aleph_1$ witnessing that a given partial map cannot be extended to an isomorphism between given quotient structures (see [15]).

Triviality of isomorphisms between Borel quotient structures is usually proved using modifications of Shelah's technique. However, in [23] a different method was used, building on [48] to prove that the assertion 'all isomorphisms between quotients $\mathcal{P}(\mathbb{N})/I$ over Borel ideal are topologically trivial' is relatively consistent with ZFC. It is not known whether this conclusion follows from forcing axioms or from Woodin's (*), and we therefore have a test question for Conjecture 5.2.

An another Π_1^2 statement with similar flavour is 'Lebesgue measure algebra does not have a Borel lifting.' Continuum Hypothesis implies the negation, and Shelah proved that this is relatively consistent with ZFC ([46]). It is, however, not known whether forcing axioms imply this conclusion.

Acknowledgments

This paper is partly based on my talks at the 'Truth and Infinity' workshop (IMS, 2011) and the 'Connes Embedding Problem' workshop (Ottawa, 2008). I would like to thank the organizers of both meetings. Another driving force for this paper—and much of my work—originated in conversations with functional analysts, too numerous to list here, over the past several years. I owe my thanks to all of them.

Appendix

I will attempt to strike a balance between formality and clarity. In particular, I will neither specify the language nor provide a complete recursive definition of a formula. In many of the instances considered below the language of ZFC is tacitly expanded to include operations or relations of the structure in question.

A.1. *Hereditary sets*

A set X is *transitive* if every element of X is a subset of X. A closure argument shows that every set is a subset of a transitive set. The minimal such set is the *transitive closure of X*. If κ is a cardinal then $H(\kappa)$ denotes the set of all sets whose transitive closure has cardinality $< \kappa$. In this paper we consider four structures of this form. $H(\aleph_0)$ (or HF), the set of hereditarily finite sets, $H(\aleph_1)$ (or HC), the set of hereditarily countable sets, $H(\aleph_2)$ and $H(\mathfrak{c}^+)$ (here $\mathfrak{c}^+ = (2^{\aleph_0})^+$, the least cardinal greater than the continuum). A simple coding argument shows that $(H(\aleph_0), \in)$ is bi-interpretable with $(\mathbb{N}, +, \cdot, 0)$ and that $(H(\aleph_1), \in)$ is bi-interpretable with $(\mathcal{P}(\mathbb{N}), \mathbb{N}, \in)$.

A.2. *Arithmetical formulas*

Arithmetical formulas are the ones in which all quantifiers range over $\mathbb{N}$. Since there is a recursive bijection between $\mathbb{N}$ and $H(\aleph_0)$, such formulas are provably equivalent to ones in which all quantifiers range over $H(\aleph_0)$. A quantification is bounded if it is of the form $(\exists x \leq n)$ or $(\forall x \leq n)$ (or $(\exists x \in y)$ and $(\forall x \in y)$, when considering x and y in $H(\aleph_0)$). An arithmetical formula is Σ_0^0 and Π_0^0 if it involves only bounded quantifiers. An arithmetical formula is Σ_{n+1}^0 if it is of the form $(\exists x_1)(\exists x_2) \ldots (\exists x_k)\phi$, where ϕ is a Π_n^0 formula. An arithmetical formula is Π_{n+1}^0 if it is of the form $(\forall x_1)(\forall x_2) \ldots (\forall x_k)\phi$, where ϕ is a Σ_n^0 formula. In both cases a block of quantifiers of the same type can be replaced by a single quantifier by using some fixed coding of finite sequences of natural numbers by natural numbers.

For example, 'm is a prime number' or 'm is a sum of two squares' are both Π_0^0 and Σ_0^0, hence Goldbach's conjecture is Π_1^0 and Twin Prime Conjecture is Π_2^0. One can expand the language and consider formulas that are, provably in ZFC, equivalent to arithmetical formulas. In particular, quantification over any fixed countable set (such as $\mathbb{Q}$, $\mathrm{SL}_2(\mathbb{Z})$, $M_n(\mathbb{Q}+i\mathbb{Q})$ $\ldots$), counts as quantification over $\mathbb{N}$.

A.3. *Analytical formulas*

In this context quantification is bounded if it is of the form $(\exists x \in \mathbb{N})$ or $(\forall x \in \mathbb{N})$. An analytical formula is Σ_0^1 and Π_0^1 if it involves only bounded quantification. Quantification in analytical formulas is over the real numbers. An analytical formula is Σ_{n+1}^1 if it is of the form $(\exists x_1)(\exists x_2) \ldots (\exists x_k)\phi$,

ϕ is a Π_n^1 formula. An analytical formula is Π_{n+1}^1 if it is of the form $(\forall x_1)(\forall x_2)\dots(\forall x_k)\phi$, where ϕ is a Σ_n^0 formula. Like in the case of arithmetic formulas, in both cases a block of quantifiers of the same type can be replaced by a single quantifier. Also, quantification over $\mathbb{N}$ (or any fixed countable set) can be absorbed into quantification over $\mathbb{R}$ without increasing the complexity.

Since any two uncountable standard Borel spaces are Borel-isomorphic, by expanding the language by a predicate for the Borel isomorphism, it is no loss of generality to allow quantification over any standard Borel space instead of $\mathbb{R}$.

A.4. *Examples*

I list some well-known problems that are equivalent to absolute formulas, and therefore unlikely to be independent from ZFC. The choice of problems is obviously biased towards my own research interests, but they should provide some idea of the concepts. Moreover, a standard absoluteness argument shows that if any of these statements can be forced over a model M of ZFC then it is already true in M. This for example implies that if CH is used to prove some of these statements then the use of CH can be removed from the proof.

Free Group Factors The assertion that the free group factors $L(F_2)$ and $L(F_3)$ are isomorphic is a Σ_1^1 statement. This is because the isomorphism is coded by a homeomorphism between two Polish spaces (e.g., the unit balls equipped with the ℓ_2-metric) which is also an algebraic isomorphism. Such a homeomorphism can be coded by a real number, and checking that it is a homeomorphism does not require quantification over uncountable sets.

Invariant Subspace Problem This is the assertion that every bounded linear operator on a separable Hilbert space has a nontrivial invariant subspace. It is easily seen to be Π_2^1.

Riemann's Hypothesis asserts that all zeros of a continuous function belong to a closed set. It is a Π_1^1 statement. It is, however, known to be equivalent to a Π_1^0 statement (essentially because the of Riemann's ζ function zeros can be enumerated).

Each of the assertions that every countable group is sofic, every countable group is hyperfinite, or every countable hyperfinite group is sofic, ([43]) is a Π_1^1 statement. This is because expressing that a group is sofic (or hyperfinite) requires only quantification over countable sets: finite subsets of the group, rationals, and finite permutation groups.

Kadison–Singer problem[d] is in its original formulation a Π_1^2 statement but it is known to be equivalent to Anderson's paving conjecture (see [8]). The latter asserts that for every $\varepsilon > 0$ there exists $k \in \mathbb{N}$ such that every n and every $n \times n$ matrix A over $\mathbb{C}$ with all diagonal entries equal to zero there exist projections p_j, for $1 \leq j \leq k$, spanned by basic vectors in $\ell_2(k)$, such that for each j we have $\|p_j A p_j\| < \varepsilon \|A\|$ (the operator norm). While the statement is, as stated, Π_3^1, it is equivalent to a Π_3^0 statement. This is because ε can be taken to be a rational, and A can be taken to be a matrix over the countable field $\mathbb{Q} + i\mathbb{Q}$.

A problem superfluously similar to the Kadison–Singer problem is Anderson's conjecture (see below).

Connes Embedding Problem states that every II_1 factor with separable predual can be embedded into an ultrapower (associated with a nonprincipal ultrafilter on $\mathbb{N}$) of the hyperfinite II_1 factor (the latter object is usually denoted by R^ω, but set theorists may want to take note that R does not stand for $\mathbb{R}$ and ω is not ω). While this formulation involves quantification over ultrafilters on $\mathbb{N}$, it is well-known that the choice of the ultafilter is irrelevant. Moreover, CEP is equivalent to the assertion that all 'microstates' computed in an arbitrary II_1 factor can be approximated by microstates computed in sufficiently large matrix algebras ([42]). For a fixed II_1 factor this is an arithmetic (more precisely, Π_2^0) statement and therefore CEP is equivalent to a Π_1^1 statement.

I now proceed to discuss some open problems that do not have a known reformulation as an absolute statement. They are more likely candidates for statements independent from ZFC than the above.

The statement of *Brown–Douglas–Fillmore problem* ([7]) whether there *exists* a K-theory reversing automorphism of the Calkin algebra is Σ_1^2. Similarly, statement that two fixed Borel quotient structures (as in §5.1) are isomorphic is Σ_1^2. The negative answer to the Brown–Douglas–Fillmore question is relatively consistent with ZFC. This is because by a simple Fredholm index argument such an automorphism cannot be inner and by [18] all automorphisms of the Calkin algebra are consistently inner.

Anderson's conjecture ([2]), that every pure state of the Calkin algebra can be diagonalized, is a Π_1^2 statement. It is known that its negation follows from CH ([2]) as well as some weaker statements ([24]).

---

[d] Added in proof: An affirmative answer to the Kadison–Singer problem has been recently given in Marcus, Adam, Daniel A. Spielman, and Nikhil Srivastava, *Interlacing families II: Mixed characteristic polynomials and the Kadison–Singer problem*, arXiv:1306.3969 (2013).

Unlike all of the above problems, *Naimark's problem* ([1]) is not a statement of $H(\mathfrak{c}^+)$, or $H(\kappa)$ for any fixed κ. It asserts that all C*-algebras have certain property, and therefore requires unbounded quantification over the universe. By [1] a negative answer to Naimark's problem follows from Jensen's $\diamondsuit_{\omega_1}$. It is not known whether a positive answer to Naimark's problem is relatively consistent with ZFC or whether $\diamondsuit_\kappa$ for some $\kappa > \aleph_1$ implies the existence of a counterexample. At present the minimal cardinality of a purported counterexample can be bounded only by using some of the strongest known large cardinal axioms (see [5]).

References

1. C. Akemann and N. Weaver, *Consistency of a counterexample to Naimark's problem*, Proc. Natl. Acad. Sci. USA **101** (2004), no. 20, 7522–7525.
2. _____, $\mathcal{B}(H)$ *has a pure state that is not multiplicative on any masa*, Proc. Natl. Acad. Sci. USA **105** (2008), no. 14, 5313–5314.
3. V.I. Arnold, *Polymathematics: is mathematics a single science or a set of arts?*, Mathematics: frontiers and perspectives, Amer. Math. Soc., Providence, RI, 2000, pp. 403–416.
4. J. Bagaria, *Bounded forcing axioms as principles of generic absoluteness*, Arch. Math. Logic **39** (2000), no. 6, 393–401. MR MR1773776 (2001i:03103).
5. _____, *C(n)-cardinals*, Arch. Math. Logic **51** (2012), no. 3-4, 213–240.
6. B. Blackadar, *Operator algebras*, Encyclopaedia of Mathematical Sciences, vol. 122, Springer-Verlag, Berlin, 2006, Theory of C^*-algebras and von Neumann algebras, Operator Algebras and Non-commutative Geometry, III.
7. L.G. Brown, R.G. Douglas, and P.A. Fillmore, *Extensions of C^*-algebras and K-homology*, Annals of Math. **105** (1977), 265–324.
8. P.G. Casazza and J.C. Tremain, *The Kadison-Singer problem in mathematics and engineering*, Proc. Natl. Acad. Sci. USA **103** (2006), no. 7, 2032–2039 (electronic).
9. C.C. Chang and H.J. Keisler, *Model theory*, third ed., Studies in Logic and the Foundations of Mathematics, vol. 73, North-Holland Publishing Co., Amsterdam, 1990.
10. S. Coskey and I. Farah, *Automorphisms of corona algebras and group cohomology*, Trans. Amer. Math. Soc. (to appear).
11. Philipp Doebler and Ralf Schindler, Π_2 *consequences of BMM+ NS_{ω_1} is precipitous and the semiproperness of stationary set preserving forcings*, Math. Res. Lett. **16** (2009), no. 5, 797–815.
12. I. Farah, *Analytic quotients: theory of liftings for quotients over analytic ideals on the integers*, Memoirs of the American Mathematical Society, vol. 148, no. 702, 2000.
13. _____, *Liftings of homomorphisms between quotient structures and Ulam stability*, Logic Colloquium '98, Lecture notes in logic, vol. 13, A.K. Peters, 2000, pp. 173–196.

14. _____, *Dimension phenomena associated with* $\beta\mathbb{N}$-*spaces*, Top. Appl. **125** (2002), 279–297.
15. _____, *Luzin gaps*, Trans. Amer. Math. Soc. **356** (2004), 2197–2239.
16. _____, *Rigidity conjectures*, Logic Colloquium 2000, Lect. Notes Log., vol. 19, Assoc. Symbol. Logic, Urbana, IL, 2005, pp. 252–271.
17. _____, *All automorphisms of all Calkin algebras*, Math. Research Letters **18** (2011), 489–503.
18. _____, *All automorphisms of the Calkin algebra are inner*, Annals of Mathematics **173** (2011), 619–661.
19. _____, *The extender algebra and* Σ_1^2-*absoluteness*, Cabal Seminar (A.S. Kechris, B. Löwe, and J. Steel, eds.), vol. IV, to appear.
20. I. Farah, R. Ketchersid, P.B. Larson, and M. Magidor, *Absoluteness for universally Baire sets and the uncountable II*, Computational Prospects of Infinity, part II (C. Chong et al., eds.), World Scientific, 2008, pp. 163–192.
21. I. Farah and P.B. Larson, *Absoluteness for universally Baire sets and the uncountable. I*, Set theory: recent trends and applications, Quad. Mat., vol. 17, Dept. Math., Seconda Univ. Napoli, Caserta, 2006, pp. 47–92.
22. I. Farah and P. McKenney, *Homeomorphisms of Cech-Stone remainders: The zero-dimensional case*, arXiv:1211.4765 (2012).
23. I. Farah and S. Shelah, *Trivial automorphisms*, Israel J. Math. (to appear).
24. I. Farah and E. Wofsey, *Set theory and operator algebras*, Appalachian set theory 2006-2010 (J. Cummings and E. Schimmerling, eds.), Cambridge University Press, 2013, pp. 63–120.
25. Solomon Feferman, Harvey M. Friedman, Penelope Maddy, and John R. Steel, *Does mathematics need new axioms?*, Bull. Symbolic Logic **6** (2000), no. 4, 401–446.
26. Q. Feng, M. Magidor, and W.H. Woodin, *Universally Baire sets of reals*, Set Theory of the Continuum (H. Judah, W. Just, and W.H. Woodin, eds.), Springer-Verlag, 1992, pp. 203–242.
27. M. Foreman, M. Magidor, and S. Shelah, *Martin's maximum, saturated ideals and nonregular ultrafilters, I*, Annals of Mathematics **127** (1988), 1–47.
28. H. Friedman, *Boolean relation theory and incompleteness*, to appear.
29. A. Kanamori, *The higher infinite: large cardinals in set theory from their beginnings*, Perspectives in Mathematical Logic, Springer, Berlin–Heidelberg–New York, 1995.
30. V. Kanovei and M. Reeken, *On Ulam's problem concerning the stability of approximate homomorphisms*, Tr. Mat. Inst. Steklova **231** (2000), 249–283.
31. _____, *New Radon–Nikodym ideals*, Mathematika **47** (2002), 219–227.
32. A.S. Kechris, *Classical descriptive set theory*, Graduate texts in mathematics, vol. 156, Springer, 1995.
33. H.J. Keisler, *Logic with the quantifier "There exists uncountably many"*, Annals of Mathematical Logic **1** (1970), 1–93.
34. R. Ketchersid, P.B. Larson, and J. Zapletal, *Regular embeddings of the stationary tower and Woodin's* Σ_2^2 *maximality theorem*, J. Symbolic Logic **75** (2010), 711–727.
35. P. Koellner, *Large cardinals and determinacy*, Stanford Encyclopaedia of Philosophy (to appear).

36. P.B. Larson, *The stationary tower*, University Lecture Series, vol. 32, American Mathematical Society, Providence, RI, 2004, Notes on a course by W.H.Woodin.

37. _____, *Martin's maximum and definability in* $H(\aleph_2)$, Ann. Pure Appl. Logic **156** (2008), no. 1, 110–122.

38. Azriel Lévy, *A hierarchy of formulas in set theory*, Mem. Amer. Math. Soc. No. **57** (1965), 76.

39. Menachem Magidor and Jerome Malitz, *Compact extensions of L(Q). Ia*, Ann. Math. Logic **11** (1977), no. 2, 217–261. MR MR0453484 (56 #11746)

40. Y.N. Moschovakis, *Descriptive set theory*, Studies in logic and foundations of mathematics, vol. 100, North–Holland, 1980.

41. I. Neeman, *Inner models in the region of a Woodin limit of Woodin cardinals*, Ann. Pure Appl. Logic **116** (2002), no. 1-3, 67–155.

42. Narutaka Ozawa, *About the QWEP conjecture*, Internat. J. Math. **15** (2004), no. 5, 501–530.

43. V. Pestov, *Hyperlinear and sofic groups: a brief guide*, Bull. Symbolic Logic **14** (2008), 449–480.

44. N.C. Phillips and N. Weaver, *The Calkin algebra has outer automorphisms*, Duke Math. Journal **139** (2007), 185–202.

45. G. Sargsyan, *Descriptive inner model theory*, Bull. Symb. Logic **19** (2013), no. 1, 1–55.

46. S. Shelah, *Proper forcing*, Lecture Notes in Mathematics, 940, Springer, 1982.

47. S. Shelah and J. Steprāns, *PFA implies all automorphisms are trivial*, Proceedings of the American Mathematical Society **104** (1988), 1220–1225.

48. _____, *Martin's axiom is consistent with the existence of nowhere trivial automorphisms*, Proc. Amer. Math. Soc. **130** (2002), no. 7, 2097–2106.

49. R. Solovay, *A model of set theory in which every set of reals is Lebesgue measurable*, Annals of Mathematics **92** (1970), 1–56.

50. John R. Steel, *What is ... a Woodin cardinal?*, Notices Amer. Math. Soc. **54** (2007), no. 9, 1146–1147.

51. B. Veličković, *OCA and automorphisms of* $\mathcal{P}(\omega)/\mathrm{Fin}$, Top. Appl. **49** (1992), 1–13.

52. M. Viale, *Martin's Maximum revisited*, preprint, arXiv:1110.1181, 2011.

53. M. Viale and C. Weiß, *On the consistency strength of the proper forcing axiom*, Adv. Math. **228** (2011), no. 5, 2672–2687.

54. John von Neumann, *The mathematician*, The Works of the Mind, The University of Chicago Press, Chicago, Ill., 1947, Edited for the Committee on Social Thought by Robert B. Heywood, pp. 180–196.

55. N. Weaver, *Set theory and* C^**-algebras*, Bull. Symb. Logic **13** (2007), 1–20.

56. W.H. Woodin, Σ_1^2*-absoluteness*, handwritten note of May 1985, 1985.

57. _____, *The continuum hypothesis. I*, Notices Amer. Math. Soc. **48** (2001), no. 6, 567–576.

58. _____, *The axiom of determinacy, forcing axioms, and the nonstationary ideal*, revised ed., de Gruyter Series in Logic and its Applications, vol. 1, Walter de Gruyter GmbH & Co. KG, Berlin, 2010.

59. _____, *Suitable extender models I*, J. Math. Log. **10** (2010), no. 1-2, 101–339.

A MULTIVERSE PERSPECTIVE ON THE AXIOM OF CONSTRUCTIBILITY[a]

Joel David Hamkins

Visiting Professor of Philosophy, New York University, USA
Professor of Mathematics, The City University of New York
The Graduate Center & College of Staten Island, USA
jhamkins@gc.cuny.edu
http://jdh.hamkins.org

I shall argue that the commonly held $V \neq L$ via maximize position, which rejects the axiom of constructibility $V = L$ on the basis that it is restrictive, implicitly takes a stand in the pluralist debate in the philosophy of set theory by presuming an absolute background concept of ordinal. The argument appears to lose its force, in contrast, on an upwardly extensible concept of set, in light of the various facts showing that models of set theory generally have extensions to models of $V = L$ inside larger set-theoretic universes.

1. Introduction

Set theorists often argue against the axiom of constructibility $V = L$ on the basis that it is restrictive. Some argue that we have no reason to think that every set should be constructible, or as Shelah puts it, "Why the hell should it be true?" [20]. To suppose that every set is constructible is seen as an artificial limitation on set-theoretic possibility, and perhaps it is a mistaken principle generally to suppose that all structure is definable. Furthermore,

---

[a]This article expands on an argument that I made during my talk at the Asian Initiative for Infinity: Workshop on Infinity and Truth, held in July 25–29, 2011 at the Institute for Mathematical Sciences, National University of Singapore. This work was undertaken during my subsequent visit at NYU in Summer and Fall, 2011, and completed when I returned to CUNY. My research has been supported in part by NSF grant DMS-0800762, PSC-CUNY grant 64732-00-42 and Simons Foundation grant 209252. Commentary concerning this paper can be made at http://jdh.hamkins.org/multiverse-perspective-on-constructibility.

although $V = L$ settles many set-theoretic questions, it seems so often to settle them in the 'wrong' way, without the elegant smoothness and unifying vision of competing theories, such as the situation of descriptive set theory under $V = L$ in comparison with that under projective determinacy. As a result, the constructible universe becomes a pathological land of counterexamples. That is bad news, but it could be overlooked, in my opinion, were it not for the much worse related news that $V = L$ is inconsistent with all the strongest large cardinal axioms. The boundary between those large cardinals that can exist in L and those that cannot is the threshold of set-theoretic strength, the entryway to the upper realm of infinity. Since the $V = L$ hypothesis is inconsistent with the largest large cardinals, it blocks access to that realm, and this is perceived as intolerably limiting. This incompatibility, I believe, rather than any issue of definabilism or descriptive set-theoretic consequentialism, is the source of the most strident end-of-the-line deal-breaking objections to the axiom of constructibility. Set theorists simply cannot accept an axiom that prevents access to their best and strongest theories, the large cardinal hypotheses, which encapsulate their dreams of what our set theory can achieve and express.

Maddy [14, 15] articulates the grounds that mathematicians often use in reaching this conclusion, mentioning especially the *maximize* maxim, saying "the view that $V = L$ contradicts *maximize* is widespread," citing Drake, Moschovakis and Scott. Steel argues that "$V = L$ is restrictive, in that adopting it limits the interpretative power of our language." He points out that the large cardinal set theorist can still understand the $V = L$ believer by means of the translation $\varphi \mapsto \varphi^L$, but "there is no translation in the other direction" and that "adding $V = L$ ... just prevents us from asking as many questions!" [21]. At bottom, the axiom of constructibility appears to be incompatible with strength in our set theory, and since we would like to study this strength, we reject the axiom.

Let me refer to this general line of reasoning as the $V \neq L$ *via maximize* argument. The thesis of this article is that the $V \neq L$ via maximize argument relies on a singularist as opposed to pluralist stand on the question whether there is an absolute background concept of ordinal, that is, whether the ordinals can be viewed as forming a unique completed totality. The argument, therefore, implicitly takes sides in the universe versus multiverse debate, and I shall argue that without that stand, the $V \neq L$ via maximize argument lacks force.

In [17, 16], Maddy gives the $V \neq L$ via maximize argument sturdier legs, fleshing out a more detailed mathematical account of it, based on a methodology of mathematical naturalism and using the idea that maximization involves realizing more isomorphism types. She begins with the 'crude version' of the argument:

> The idea is simply this: there are things like $0^\sharp$ that are not in L. And not only is $0^\sharp$ not in L; its existence implies the existence of an isomorphism type that is not realized by anything in L. ... So it seems that $\mathsf{ZFC} + V = L$ is restrictive because it rules out the extra isomorphism types available from $\mathsf{ZFC} + \exists 0^\sharp$. [17]

For the full-blown argument, she introduces the concept of a 'fair interpretation' of one theory in another and the idea of one theory maximizing over another, leading eventually to a proposal of what it means for a theory to be 'restrictive' (see the details in Section 2), showing that $\mathsf{ZFC} + V = L$ and other theories are restrictive, as expected, in that sense.

My thesis in this article is that the general line of the $V \neq L$ via maximize argument presumes that we have an absolute background concept of ordinal, that the ordinals build up to form an absolute completed totality. Of course, many set-theorists do take that stand, particularly set theorists in the California school. The view that the ordinals form an absolute completed totality follows, of course, from the closely related view that there is a unique absolute background concept of set, by which the sets accumulate to form the entire set-theoretic universe V, in which every set-theoretic assertion has a definitive final truth value. Martin essentially argues for the equivalence of these two commitments in his categoricity argument [18], where he argues for the uniqueness of the set-theoretic universe, an argument that is a modern-day version of Zermelo's categoricity argument with strong parallels in Isaacson's [11]. Martin's argument is founded on the idea of an absolute unending well-ordered sequence of set-formation stages, an 'Absolute Infinity' as with Cantor. Although Martin admits that 'it is of course possible to have doubts about the sharpness of the concept of wellordering," [18], his argument presumes that the concept is sharp, just as I claim the $V \neq L$ via maximize argument does.

Let me briefly summarize the position I am defending in this article, which I shall describe more fully in Section 4. On the upwardly extensible concept of set, one holds that any given concept of set or set-theoretic universe may always be extended to a much better one, with more sets and larger ordinals. Perhaps the original universe even becomes a mere countable set in the extended universe. The 'class of all ordinals', on this view, makes sense only relative to a particular set-theoretic universe, for there is no expectation that these extensions cohere or converge. This multiverse perspective resonates with or even follows from a higher-order version of the maximize principle, where we maximize not merely which sets exist, but also which set-theoretic universes exist. Specifically, it would be limiting for one set-theoretic universe to have all the ordinals, when we can imagine another universe looking upon it as countable. Maximize thereby leads us to expect that every set-theoretic universe should not only have extensions, but extremely rich extensions, satisfying extremely strong theories, with a full range of large cardinals. Meanwhile, I shall argue, the mathematical results of Section 3 lead naturally to the additional conclusion that every set-theoretic universe should also have extensions satisfying $V = L$. In particular, even if we have very strong large cardinal axioms in our current set-theoretic universe V, there is a much larger universe V^+ in which the former universe V is a countable transitive set and the axiom of constructibility holds. This perspective, by accommodating both large cardinals and $V = L$ in the multiverse, appears to dissolve the principal thrust of the $V \neq L$ via maximize argument. The idea that $V = L$ is permanently incompatible with large cardinals evaporates when we can have large cardinals and reattain $V = L$ in a larger domain. In this way, $V = L$ no longer seems restrictive, and the upward extensible concept of set reveals how large cardinals and other strong theories, as well as $V = L$, may all be pervasive as one moves up in the multiverse.

2. Some New Problems with Maddy's Proposal

Although my main argument is concerned only with the general line of the $V \neq L$ via maximize position, rather than with Maddy's much more specific account of it in [17], before continuing with my main agument I would nevertheless like to mention a few problems with that specific proposal.

To quickly summarize the details, she defines that a theory T *shows φ is an inner model* if T proves that φ defines a transitive class satisfying every instance of an axiom of ZFC, and either T proves every ordinal is in the class, or T proves that there is an inaccessible cardinal κ, such that every ordinal less than κ is in the class. Next, φ is a *fair interpretation* of T in T', where T extends ZFC, if T' shows φ is an inner model and T' proves every axiom of T for this inner model. A theory T' *maximizes* over T, if there is a fair interpretation φ of T in T', and T' proves that this inner model is not everything (let's assume T' includes ZFC). The theory T' *properly maximizes* over T if it maximizes over T, but not conversely. The theory

T' *strongly maximizes* over T if the theories contradict one another, T' maximizes over T and no consistent extension T'' of T properly maximizes over T'. All of this culminates in her final proposal, which is to say that a theory T is *restrictive* if and only if there is a consistent theory T' that strongly maximizes over it.

Let me begin with a quibble concerning the syntactic form of her definition of 'shows φ is an inner model', which in effect requires T to settle the question of whether the inner model is to contain all ordinals or instead merely all ordinals up to an inaccessible cardinal. That is, she requires that either T proves that φ is in the first case or that T proves that φ is in the second case, rather than the weaker requirement that T prove merely that φ is in one of the two cases (so the distinction is between $(T \vdash A) \vee (T \vdash B)$ and $T \vdash A \vee B$). To illustrate how this distinction plays out in her proposal, consider the theory Inacc = ZFC + 'there are unboundedly many inaccessible cardinals' and the theory $T =$ ZFC + 'either there is a Mahlo cardinal or there are unboundedly many inaccessible cardinals in L.' (I shall assume without further remark that these large cardinal theories and the others I mention are consistent.) Every model of T has an inner model of Inacc, either by truncating at the Mahlo cardinal, if there is one, or by going to L, if there is not. Thus, we seem to have inner models of the form Maddy desires. Unfortunately, however, this is not good enough, and I claim that Inacc is actually *not* fairly interpreted in T. To see this, notice first that T does not prove the existence of an inaccessible cardinal, since we can force over any model of Inacc by destroying all inaccessible cardinals and thereby produce a model of T having no inaccessible cardinals.[a] Consequently, if T shows φ is an inner model, it cannot be because of the second clause, which requires T to prove the existence of an inaccessible cardinal. Thus, T must prove φ holds of all ordinals. But notice also that T does not prove that there are unboundedly many inaccessible cardinals in L, since by truncation we can easily have a Mahlo cardinal in L with no inaccessible cardinals above it. So T also cannot prove that φ defines a proper class model of Inacc. Thus, Inacc is not fairly interpreted in T, even though we might have wished it to be. This issue can be addressed, of course, by modifying the definition of shows-an-inner-model to subsume the disjunction under the provability sign, that is, by requiring instead that T prove the disjunction that either

---

[a]First force 'Ord is not Mahlo' by adding a closed unbounded class C of non-inaccessible cardinals—this forcing adds no new sets—and then perform Easton forcing to ensure $2^\gamma = \delta^+$ whenever γ is regular and δ is the next element of C.

φ holds of all ordinals or that it holds of all ordinals up to an inaccessible cardinal. But let me leave this issue; it does not affect my later comments.

My next objection is that the fairly-interpreted-in relation is not transitive, whereas our pre-reflective ideas for an interpreted-in relation would call for it to be transitive. That is, I claim that it can happen that a first theory has a fair interpretation in a second, which has a fair interpretation in a third, but the first theory has no fair interpretation in the third. Here is a specific example showing the lack of transitivity:

$$R = \mathsf{ZFC} + V = L + \text{there is no inaccessible cardinal,}$$
$$S = \mathsf{ZFC} + V = L + \text{there is an inaccessible cardinal,}$$
$$T = \mathsf{ZFC} + \omega_1 \text{ is inaccessible in } L.$$

The reader may easily verify that R has a fair interpretation in S by truncating the universe at the first inaccessible cardinal, and S has a fair interpretation in T by going to L. Furthermore, every model of S has forcing extensions satisfying T, by the Lévy collapse. Meanwhile, I claim that R has no fair interpretation in T. The reason is that T is consistent with the lack of inaccessible cardinals, and so if T shows φ is an inner model, then in any model of T having no inaccessible cardinals, this inner model must contain all the ordinals. In this case, in order for it to have R^φ, the inner model must be all of L, which according to T has an inaccessible cardinal, and therefore does not satisfy R after all. So R is not fairly interpreted in T. The reader may construct many similar examples of intransitivity. The essence here is that the first theory is fairly interpreted in the second only by truncating, and the second is fairly interpreted in the third only by going to an inner model containing all the ordinals, but there is no way to interpret the first in the third except by doing both, which is not allowed in the definition if the truncation point is inaccessible only in the inner model and not in the larger universe.

The same example shows that the maximizing-over relation also is not transitive, since T maximizes over S and S maximizes over R, by the fair interpretations mentioned above (note that these theories are mutually exclusive), but T does not maximize over R, since R has no fair interpretation in T. Similarly, the reader may verify that the example shows that the properly-maximizes-over and the strongly-maximizes-over relations also are not transitive.

Let me turn now to give a few additional examples of what Maddy calls a 'false positive,' a theory deemed formally restrictive, which we do not find intuitively to be restrictive. As I see it, the main purpose of [17] is to give precise mathematical substance to the intuitive idea that some set theories seem restrictive in a way that others do not. We view $V = L$ and 'there is a largest inaccessible cardinal' as limiting, while 'there are unboundedly many inaccessible cardinals' seems open-ended and unrestrictive. Maddy presents some false positives, including an example of Steel's showing that ZFC + 'there is a measurable cardinal' is restrictive because it is strongly maximized by the theory ZFC + $0^\dagger$ exists + $\forall \alpha < \omega_1 \ L_\alpha[0^\dagger] \not\models$ ZFC. Löwe points out that "this example can be generalized to at least every interesting theory in large cardinal form extending ZFC. Thus, most theories are restrictive in a formal sense," [12] and he shows in [13] that ZFC itself is formally restrictive because it is maximized by the theory ZF + 'every uncountable cardinal is singular'.

I would like to present examples of a different type, which involve what I believe to be more attractive maximizing theories that seem to avoid the counterarguments that have been made to the previous examples of false positives. First, consider again the theory Inacc, asserting ZFC + 'there are unboundedly many inaccessible cardinals', a theory Maddy wants to regard as not restrictive. Let T be the theory asserting ZFC + 'there are unboundedly many inaccessible cardinals in L, but no worldly cardinals in V.' A cardinal κ is *worldly* when $V_\kappa \models$ ZFC. Worldliness is a weakening of inaccessibility, since every inaccessible cardinal is worldly and in fact a limit of worldly cardinals; but meanwhile, worldly cardinals need not be regular, and the regular worldly cardinals are exactly the inaccessible cardinals. The worldly cardinals often serve as a substitute for inaccessible cardinals, allowing one to weaken the large cardinal commitment of a hypothesis. For example, one may carry out most uses of the Grothendieck universe axiom in category theory by using mere worldly cardinals in place of inaccessible cardinals. The theory Inacc has a fair interpretation in T, by going to L, and as a result, T maximizes over Inacc. Meanwhile, I claim that no strengthening of Inacc properly maximizes over T. To see this, suppose that Inacc$^+$ contains Inacc and shows φ is an inner model M satisfying T. If M contains all the ordinals, then since Inacc proves that the inaccessible cardinals are unbounded, M would have to contain all those inaccessible cardinals, which would remain inaccessible in M since inaccessibility is downward absolute, and therefore violate the claim of T that there are no worldly cardinals. So by the definition of fair interpretation, therefore, M would have to contain all the ordinals up to an inaccessible cardinal κ. But in this case, a Loweheim-Skolem argument shows that there is a closed unbounded set of $\gamma < \kappa$ with $V_\gamma^M \prec V_\kappa^M$, and all such γ would be worldly cardinals in

M, violating T. Thus, Inacc is strongly maximized by T, and so Inacc is restrictive.

Let me improve the example to make it more attractive, provided that we read Maddy's definition of 'fair interpretation' in a way that I believe she may have intended. The issue is that although Maddy refers to 'truncation... at inaccessible levels' and her definition is typically described by others using that phrase, nevertheless the particular way that she wrote her definition does not actually ensure that the truncation occurs at an inaccessible level. Specifically, in the truncation case, she writes that T should prove that there is an inaccessible cardinal κ for which $\forall\alpha(\alpha < \kappa \to \varphi(\alpha))$. But should this implication be a biconditional? Otherwise, of course, nothing prevents φ from continuing past κ, and the definition would be more accurately described as 'truncation at, or somewhere above, an inaccessible cardinal'. If one wants to allow truncation at non-inaccessible cardinals, why should we bother to insist that the height should exceed some inaccessible cardinal? Replacing this implication with a biconditional would indeed ensure that when the inner model arises by truncation, it does so by truncating at an inaccessible cardinal level. So let us modify the reading of 'fair interpretation' so that truncation, if it occurs, does so at an inaccessible cardinal level. In this case, consider the theory Inacc as before, and let MC* be the theory ZFC + 'there is a measurable cardinal with no worldly cardinals above it'. By truncating at a measurable cardinal, we produce a model of Inacc, and so MC* offers a fair interpretation of Inacc, and consequently MC* maximizes over Inacc. But no consistent strengthening of Inacc can maximize over MC*, since if $V \models$ Inacc and W is an inner model of V satisfying MC*, then W cannot contain all the ordinals of V, since the inaccessible cardinals would be worldly in W, and neither can the height of W be inaccessible in V, since if $\kappa = W \cap \mathrm{Ord}$ is inaccessible in V, then by a Lowenheim-Skolem argument there must be a closed unbounded set of $\gamma < \kappa$ such that $W_\gamma \prec W$, and this will cause unboundedly many worldly cardinals in W, contrary to MC*. Thus, on the modified definition of fair interpretation, we conclude that MC* strongly maximizes over Inacc, and so Inacc is restricted.

One may construct similar examples using the theory ZFC + 'there is a proper class of measurable cardinals', which is strongly maximized by SC* = ZFC + 'there is a supercompact cardinal with no worldly cardinals above it'. Truncating at the supercompact cardinal produces a model of the former theory, but no strengthening of the former theory can show SC* in an inner model, since the unboundedly many measurable cardinals of the former theory prevent the showing of any proper class model of SC*, and the eventual lack of worldly cardinals in SC* prevents it from being shown in any truncation at an inaccessible level of any model of ZFC. A general format for these examples would be ZFC + 'there is a proper class of large cardinals of type LC' and $T =$ ZFC + 'there is an inaccessible limit of LC cardinals, with no worldly cardinals above.' Such examples work for any large cardinal notion LC that implies worldliness, is absolute to truncations at inaccessible levels and is consistent with a lack of worldly cardinals above. Almost all (but not all) of the standard large cardinal notions have these features.

Maddy has rejected some of the false positives on the grounds that the strongly maximizing theory involved is a 'dud' theory, such as ZFC + ¬ Con(ZFC). Are the theories above, MC* and SC*, duds in this sense? It seems hard to argue that they are. For various reasons, set theorists often consider models of set theory with largest instances of large cardinals and no large cardinals above, often obtaining such models by truncation, in order to facilitate certain constructions. Indeed, the idea of truncating the universe at an inaccessible cardinal level lies at the heart of Maddy's definitions. But much of the value of that idea is already obtained when one truncates at the worldly cardinals instead. The theory MC* can be obtained from any model of measurable cardinal by truncating at the least worldly cardinal above it, if there is one, and similarly in the case of SC*. Furthermore, since we can also often obtain MC* and SC* by moving from a large cardinal model to a forcing extension, where all the previous context and strength seems still available, these theories do not seem to be duds in any obvious way. Nevertheless, the theories MC* and SC* are restrictive, of course, in the intuitive sense that Maddy's project is concerned with. But to object that these theories are duds on the grounds that they are restrictive would be to give up the entire project; the point was to give precise substance to our notion of 'restrictive', and it would beg the question to define that a theory is restrictive if it is strongly maximized by a theory that is not 'restrictive.'

3. Several Ways in which $V = L$ is Compatible with Strength

In order to support my main thesis, I would like next to survey a series of mathematical results, most of them a part of set-theoretic folklore, which reveal various senses in which the axiom of constructibility $V = L$ is compatible with strength in set theory, particularly if one has in mind the possibility of moving from one universe of set theory to a much larger one.

First, there is the easy observation, expressed in Observation 3.1, that L and V satisfy the same consistency assertions. For any constructible theory T in any language—and by a 'constructible' theory I mean just that $T \in L$, which is true of any c.e. theory, such as ZFC plus any of the usual large cardinal hypotheses—the constructible universe L and V agree on the consistency of T because they have exactly the same proofs from T. It follows from this, by the completeness theorem, that they also have models of exactly the same constructible theories.

Observation 3.1: The constructible universe L and V agree on the consistency of any constructible theory. They have models of the same constructible theories.

What this easy fact shows, therefore, is that while asserting $V = L$ we may continue to make all the same consistency assertions, such as $\mathrm{Con}(\mathsf{ZFC} + \exists$ measurable cardinal$)$, with exactly the same confidence that we might hope to do so in V, and we correspondingly find models of our favorite strong theories inside L. Perhaps a skeptic worries that those models in L are somehow defective? Perhaps we find only ill-founded models of our strong theory in L? Not at all, in light of the following theorem, a fact that I found eye-opening when I first came to know it years ago.

Theorem 3.2: *The constructible universe L and V have transitive models of exactly the same constructible theories in the language of set theory.*

Proof: The assertion that a given theory T has a transitive model has complexity $\Sigma^1_2(T)$, in the form "there is a real coding a well founded structure satisfying T," and so it is absolute between L and V by the Shoenfield absoluteness theorem, provided the theory itself is in L.   □

Consequently, one can have transitive models of extremely strong large cardinal theories without ever leaving L. For example, if there is a transitive model of the theory ZFC + "there is a proper class of Woodin cardinals," then there is such a transitive model inside L. The theorem has the following interesting consequence.

Corollary 3.3: *(Levy-Shoenfield absoluteness theorem) In particular, L and V satisfy the same Σ_1 sentences, with parameters hereditarily countable in L. Indeed, $L_{\omega_1^L}$ and V satisfy the same such sentences.*

Proof: Since L is a transitive class, it follows that L is a Δ_0-elementary substructure of V, and so Σ_1 truth easily goes upward from L to V. Conversely, suppose V satisfies $\exists x \, \varphi(x, z)$, where φ is Δ_0 and z is hereditarily countable in L. Thus, V has a transitive model of the theory $\exists x \varphi(x, z)$, together with the atomic diagram of the transitive closure z and a bijection of it to ω. By Observation 3.1, it follows that L has such a model as well. But a transitive model of this theory in L implies that there really is an $x \in L$ with $\varphi(x, z)$, as desired. Since the witness is countable in L, we find the witness in $L_{\omega_1^L}$. $\qquad\square$

One may conversely supply a direct proof of Corollary 3.3 via the Shoenfield absoluteness theorem and then view Theorem 3.2 as the consequence, because the assertion that there is a transitive model of a given theory in L is Σ_1 assertion about that theory.

I should like now to go further. Not only do L and V have transitive models of the same strong theories, but what is more, any given model of set theory can, in principle, be continued to a model of $V = L$. Consider first the case of a countable transitive model $\langle M, \in \rangle$.

Theorem 3.4: *Every countable transitive set is a countable transitive set in the well-founded part of an ω-model of $V = L$.*

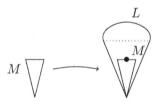

Proof: The statement is true inside L, since every countable transitive set in L is an element of some countable L_α, which is transitive and satisfies $V = L$. Further, the complexity of the assertion is Π_2^1, since it asserts that for every countable transitive set, there is another countable object satisfying a certain arithmetic property in relation to it. Consequently, by the Shoenfield absoluteness theorem, the statement is true. $\qquad\square$

Thus, every countable transitive set has an end-extension to a model of $V = L$ in which it is a set. In particular, if we have a countable transitive model $\langle M, \in \rangle \models$ ZFC, and perhaps this is a model of some very strong large cardinal theory, such as a proper class of supercompact cardinals, then nevertheless there is a model $\langle N, \in^N \rangle \models V = L$ which has M as an element, in such a way that the membership relation of $\in^N$ agrees with $\in$ on the members of M. This implies that the ordinals of N are well-founded at least to the height of M, and so not only is N an ω-model, but it is an ξ-model where $\xi = \mathrm{Ord}^M$, and we may assume that the membership relation $\in^N$ of N is the standard relation $\in$ for sets of rank up to and far exceeding ξ. Furthermore, we may additionally arrange that the model satisfies ZFC$^-$, or any desired finite fragment of ZFC, since this additional requirement is achievable in L and the assertion that it is met still has complexity Π^1_2. If there are arbitrarily large $\lambda < \omega^L_1$ with $L_\lambda \models$ ZFC, a hypothesis that follows from the existence of a single inaccessible cardinal (or merely from an uncountable transitive model of ZF), then one can similarly obtain ZFC in the desired end-extension.

A model of set theory is *pointwise definable* if every object in the model is definable there without parameters. This implies $V = \mathrm{HOD}$, since in fact no ordinal parameters are required, and one should view it as an extremely strong form of $V = \mathrm{HOD}$, although the pointwise definability property, since it implies that the model is countable, is not first-order expressible. The main theorem of [10] is that every countable model of ZFC (and similarly for GBC) has a class forcing extension that is pointwise definable.

Theorem 3.5: *If there are arbitrarily large $\lambda < \omega^L_1$ with $L_\lambda \models$ ZFC, then every countable transitive set M is a countable transitive set inside a structure M^+ that is a pointwise-definable model of ZFC $+ V = L$, and M^+ is well founded as high in the countable ordinals as desired.*

Proof: See [10] for further details. First, note that every real z in L is in a pointwise definable L_α, since otherwise, the L-least counterexample z would be definable in L_{ω_1} and hence in the Skolem hull of $\emptyset$ in L_{ω_1}, which collapses to a pointwise definable L_α in which z is definable, a contradiction. For any such α, let $L_\lambda \models$ ZFC have exactly α many smaller L_β satisfying ZFC, and so α and hence also z is definable in L_λ, whose Skolem hull of $\emptyset$ therefore collapses to a pointwise definable model of ZFC $+ V = L$ containing z. So the conclusion of the theorem is true in L. Since the complexity of this assertion is Π^1_2, it is therefore absolute to V by the Shoenfield absoluteness theorem. $\square$

Theorems 3.4 and 3.5 admit of some striking examples. Suppose for instance that $0^\sharp$ exists. Considering it as a real, the argument shows that $0^\sharp$ exists inside a pointwise definable model of ZFC $+ V = L$, well-founded far beyond ω^L_1. So we achieve the bizarre situation in which the true $0^\sharp$ sits unrecognized, yet definable, inside a model of $V = L$ which is well-founded a long way. For a second example, consider a forcing extension $V[g]$ by the forcing to collapse ω_1 to ω. The generic filter g is coded by a real, and so in $V[g]$ there is a model $M \models \mathsf{ZFC} + V = L$ with $g \in M$ and M well-founded beyond ω_1^V. The model M believes that the generic object g is actually constructible, constructed at some (necessarily nonstandard) stage beyond ω_1^V. Surely these models are unusual.

The theme of these arguments goes back, of course, to an elegant theorem of Barwise, Theorem 3.6, asserting that every countable model of ZF has an end-extension to a model of $\mathsf{ZFC} + V = L$. In Barwise's theorem, the original model becomes merely a subset of the end-extension, rather than an element of the end-extension as in Theorems 3.4 and 3.5. By giving up on the goal of making the original universe itself a set in the end-extension, Barwise seeks only to make the elements of the original universe constructible in the extension, and is thereby able to achieve the full theory of $\mathsf{ZFC} + V = L$ in the end-extension, without the extra hypothesis as in theorem 3.5, which cannot be omitted there. Another important difference is that Barwise's Theorem 3.6 also applies to nonstandard models.

Theorem 3.6: *(Barwise [2]) Every countable model of* ZF *has an end-extension to a model of* $\mathsf{ZFC} + V = L$.

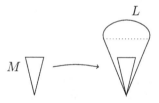

Let me briefly outline a proof in the case of a countable transitive model $M \models \mathsf{ZF}$. For such an M, let T be the theory ZFC plus the infinitary assertions $\sigma_a = \forall z\,(z \in \check{a} \iff \bigvee_{b \in a} z = \check{b})$, for every $a \in M$, in the $L_{\omega_1,\omega}$ language of set theory with constant symbol $\check{a}$ for every element $a \in M$. The σ_a assertions, which are expressible in $L_{\infty,\omega}$ logic in the sense of M, ensure that the models of T are precisely (up to isomorphism) the end-extensions of M satisfying ZFC. What we seek, therefore, is a model of the theory $T + V = L$. Suppose toward contradiction that there is none. I claim consequently that there is a proof of a contradiction from $T + V = L$ in the infinitary deduction system for $L_{\infty,\omega}$ logic, with such infinitary rules as: from σ_i for $i \in I$, deduce $\bigwedge_i \sigma_i$. Furthermore, I claim that there is such a proof inside M. Suppose not. Then M thinks that the theory $T + V = L$ is consistent in $L_{\infty,\omega}$ logic. We may therefore carry out a Henkin construction over M by building a new theory $T^+ \subseteq M$ extending $T + V = L$, with infinitely many new constant symbols, adding one new sentence at a time, each involving only finitely many of the new constants, in such a way so as to ensure that (i) the extension at each stage remains M-consistent; (ii) T^+ eventually includes any given $L_{\infty,\omega}$ sentence in M or its negation, for sentences involving only finitely many of the new constants; (iii) T^+ has the Henkin property in that it contains $\exists x\, \varphi(x, \vec{c}) \implies \varphi(d, \vec{c})$, where d is a new constant symbol used expressly for this formula; and (iv) whenever a disjunct $\bigvee_i \sigma_i$ is in T^+, then also some particular σ_i is in T^+. We may build such a T^+ in ω many steps just as in the classical Henkin construction. If N is the Henkin model derived from T^+, then an inductive argument shows that N satisfies every sentence in T^+, and in particular, it is a model of $T + V = L$, which contradicts our assumption that this theory had no model. So there must be a proof of a contradiction from $T + V = L$ in the deductive system for $L_{\infty,\omega}$ logic inside M. Since the assertion that there is such a proof is Σ_1 assertion in the language of set theory, it follows by the Levy-Shoenfield theorem (Corollary 3.3) that there is such a proof inside L^M, and indeed, inside $L_{\omega_1}^M$. This proof is a countable object in L^M and uses the axioms σ_a only for $a \in L_{\omega_1}^M$. But L^M satisfies the theory $T + V = L$ and also σ_a for all such a and hence is a model of the theory from which we had derived a contradiction. This violates soundness for the deduction system, and so $T + V = L$ has a model after all. Consequently, M has an end-extension satisfying $\mathsf{ZFC} + V = L$, as desired, and this completes the proof.

We may attain a stronger theorem, where every $a \in M$ becomes countable in the end-extension model, simply by adding the assertions '$\check{a}$ is countable' to the theory T. The point is that ultimately the proof of a contradiction exists inside $L_{\omega_1}^M$, and so the model L^M satisfies these additional assertions for the relevant a. Similarly, we may also arrange that the end-extension model is pointwise definable, meaning that every element in it is definable without parameters. This is accomplished by adding to T the infinitary assertions $\forall z \bigvee_\varphi \forall x(\varphi(x) \iff x = z)$, taking the disjunct over all first-order formulas φ. These assertions ensure that every z is defined by a first-order formula, and the point is that the σ_a arising in the proof can be taken not only from L^M, but also from amongst the definable elements of L^M, since these constitute an elementary substructure of L^M.

Remarkably, the theorem is true even for nonstandard models $\mathcal{M}$, but the proof above requires modification, since the infinitary deductions of M

may not be well-founded deductions, and this prevents the use of soundness to achieve the final contradiction. (One can internalize the contradiction to soundness, if M should happen to have an uncountable $L_\beta \models \mathsf{ZFC}$, or even merely arbitrarily large such β below $(\omega_1^L)^M$.) To achieve the general case, however, Barwise uses his compactness theorem [1] and the theory of admissible covers to replace the ill-founded model $\mathcal{M}$ with a closely related admissible set in which one may find the desired well-founded deductions and ultimately carry out an essentially similar argument. I refer the reader to the accounts in [2] and in [3].

It turns out, however, that one does not need this extra technology in the case of an ω-nonstandard model $\mathcal{M}$ of ZF, and so let me explain this case. Suppose that $\mathcal{M} = \langle M, \in^\mathcal{M} \rangle$ is an ω-nonstandard model of ZF. Let T again be the theory $\mathsf{ZFC} + \sigma_a$ for $a \in M$, where again $\sigma_a = \forall z\, (z \in \check{a} \iff \bigvee_{b \in^{\mathcal{M}} a} z = \check{b})$. Suppose there is no model of $T + V = L$. Consider the nonstandard theory ZFC^M, which includes many nonstandard formulas. By the reflection theorem, every finite collection of ZFC axioms is true in arbitrarily large $L_\beta^\mathcal{M}$, and so by overspill there must be a nonstandard finite theory ZFC^* in $\mathcal{M}$ that includes every standard ZFC axiom and which $\mathcal{M}$ believes to hold in some $L_\beta^\mathcal{M}$ for some uncountable ordinal β in $\mathcal{M}$. Let T^* be the theory ZFC^* plus all the σ_a for $a \in M$. This theory is Σ_1 definable in $\mathcal{M}$, and I claim that $\mathcal{M}$ must have a proof of a contradiction from $T^* + V = L$ in the infinitary logic $L_{\infty,\omega}^\mathcal{M}$. If not, then the same Henkin construction as above still works, working with nonstandard formulas inside $\mathcal{M}$, and the corresponding Henkin model satisfies all the actual (well-founded) assertions in $T^* + V = L$, which includes all of $T + V = L$, contradicting our preliminary assumption. So $\mathcal{M}$ has a proof of a contradiction from $T^* + V = L$. Since the assertion that there is such a proof is Σ_1, we again find a proof in L^M and even in $L_{\omega_1}^M$. But we may now appeal to the fact that M thinks L_β^M is a model of ZFC^* plus σ_a for every $a \in L_{\omega_1}^M$, which contradicts the soundness principle of the infinitary deduction system *inside* M. The point is that even though the deduction is nonstandard, this doesn't matter since we are applying soundness not externally but inside $\mathcal{M}$. The contradiction shows that $T + V = L$ must have a model after all, and so $\mathcal{M}$ has an end-extension satisfying $\mathsf{ZFC} + V = L$, as desired. Furthermore, we may also ensure that every element of $\mathcal{M}$ becomes countable in the end-extension as before.

Let me conclude this section by mentioning another sense in which every countable model of set theory is compatible in principle with $V = L$.

Theorem 3.7: *(Hamkins [6]) Every countable model of set theory* $\langle M, \in^M \rangle$, *including every transitive model, is isomorphic to a submodel of its own constructible universe* $\langle L^M, \in^M \rangle$. *In other words, there is an embedding* $j : M \to L^M$, *which is elementary for quantifier-free assertions.*

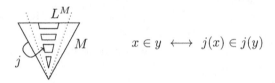

$$x \in y \;\longleftrightarrow\; j(x) \in j(y)$$

Another way to say this is that every countable model of set theory is a submodel of a model isomorphic to L^M. If we lived inside M, then by adding new sets and elements, our universe could be transformed into a copy of the constructible universe L^M.

4. An Upwardly Extensible Concept of Set

I would like now to explain how the mathematical facts identified in the previous section weaken support for the $V \neq L$ via maximize position, particularly for those set theorists inclined toward a pluralist or multiverse conception of the subject.

To my way of thinking, Theorem 3.2 already provides serious resistance to the $V \neq L$ via maximize argument, even without the multiverse ideas I shall subsequently discuss. The point is simply that much of the force and content of large cardinal set theory, presumed lost under $V = L$, is nevertheless still provided when the large cardinal theory is undertaken merely with countable transitive models, and Theorem 3.2 shows that this can be done while retaining $V = L$. We often regard a large cardinal argument or construction as important—such as Baumgartner's forcing of PFA over a model with a supercompact cardinal—because it helps us to understand a greater range for set-theoretic possibility. The fact that there is indeed an enormous range of set-theoretic possibility is the central discovery of the last half-century of set theory, and one wants a philosophical account of the phenomenon. The large cardinal arguments enlarge us by revealing the set-theoretic situations to which we might aspire. Because of the Baumgartner argument, for example, we may freely assert ZFC + PFA with the same gusto and confidence that we had for ZFC plus a supercompact cardinal, and furthermore we gain detailed knowledge about how to transform a universe of the latter theory to one of the former and how these worlds are related.[b] Modifications of that construction are what led us to worlds where MM holds and MM$^+$ and so on. From this perspective, a large part of the value of large cardinal argument is supplied already by our ability to carry it out over a transitive model of ZFC, rather than over the full universe V.

The observation that we gain genuine set-theoretic insights when working merely over countable transitive models is reinforced by the fact that the move to countable transitive models is or at least was, for many set theorists, a traditional part of the official procedure by which the forcing technique was formalized. (Perhaps a more common contemporary view is that this is an unnecessary pedagogical simplification, for one can formalize forcing over V internally as a ZFC construction.) Another supporting example is provided by the inner model hypothesis of [4], a maximality-type principle whose very formalization seems to require one to think of the universe as a toy model, for the axiom is stated about V as it exists as a countable transitive model in a larger universe. In short, much of what we hope to achieve with our strong set theories is already achieved merely by having transitive models of those theories, and Theorem 3.2 shows that the existence of any and all such kind of transitive models is fully and equally consistent with our retaining $V = L$. Because of this, the $V \neq L$ via maximize argument begins to lose its force.

Nearly every set theorist entertaining some strong set-theoretic hypothesis ψ is generally also willing to entertain the hypothesis that ZFC $+ \psi$ holds in a transitive model. To be sure, the move from a hypothesis ψ to the assertion 'there is a transitive model of ZFC $+ \psi$' is strictly increasing in consistency strength, a definite step up, but a small step. Just as philosophical logicians have often discussed the general principle that if you are willing to assert a theory T, then you are also or should also be willing to assert that 'T is consistent,' in set theory we have the similar principle, that if you are willing to assert T, then you are or should be willing to assert that 'there is a transitive model of T'. What is more, such a principle amounts essentially to the mathematical content of the philosophical reflection arguments, such as in [19], that are often used to justify large cardinal axioms. As a result, one has a kind of translation that maps any strong set-theoretic hypothesis ψ to an assertion 'there is a transitive model

---

[b]The converse question, however, whether we may transform models of PFA to models of ZFC $+ \exists$ supercompact cardinal, remains open. Many set theorists have conjectured that these theories are equiconsistent.

of ZFC + ψ', which has the same explanatory force in terms of describing
the range of set-theoretic possibility, but which because of the theorems of
Section 3 remains compatible with $V = L$.

This perspective appears to rebut Steel's claims, mentioned in the opening section of this article, that "there is no translation" from the large cardinal realm to the $V = L$ context and that "adding $V = L$ ... prevents us
from asking as many questions." Namely, the believer in $V = L$ seems fully
able to converse meaningfully with any large cardinal set theorist, simply
by imagining that the large cardinal set theorist is currently living inside a
countable transitive model. By applying the translation

$$\psi \quad \longmapsto \quad \text{'there is a transitive model of ZFC} + \psi',$$

the $V = L$ believer steps up in strength above the large cardinal set theorist,
while retaining $V = L$ and while remaining fully able to analyze and carry
out the large cardinal set theorist's arguments and constructions inside that
transitive model. Furthermore, if the large cardinal set theorist believes in
her axiom because of the philosophical reflection principle arguments, then
she agrees that set-theoretic truth is ultimately captured inside transitive
sets, and so ultimately she agrees with the step up that the $V = L$ believer
made, to put the large cardinal theory inside a transitive set. This simply
reinforces the accuracy with which the $V = L$ believer has captured the
situation.

Although the translation I am discussing is not a 'fair interpretation'
in the technical sense of [17], as discussed in Section 2, nevertheless it does
seem to me to be a fair interpretation in a sense that matters, because it
allows the $V = L$ believer to understand and appreciate the large cardinal
set theorist's arguments and constructions.

Let me now go a bit further. My claim is that on the multiverse view as I
describe it in [9] (see also [5, 8, 7]), the nature of the full outer multiverse of
V is revealed in part by the toy simulacrum of it that we find amongst the
countable models of set theory. For all we know, our current set-theoretic
universe V is merely a countable transitive set inside another much larger
universe V^+, which looks upon V as a mere toy. And so when we can prove
that a certain behavior is pervasive in the toy multiverse of any model
of set theory, then we should expect to find this behavior also in the toy
multiverse of V^+, which includes a meaningfully large part of the actual
multiverse of V. In this way, we come to learn about the full multiverse
of V by undertaking a general study of the toy model multiverses. Just
as every countable model has actual forcing extensions, we expect our full

universe to have actual forcing extensions; just as every countable model can be end-extended to a model of $V = L$, we expect the full universe V can be end-extended to a universe in which $V = L$ holds; and so on. How fortunate it is that the study of the connections between the countable models of set theory is a purely mathematical activity that can be carried out within our theory. This mathematical knowledge, such as the results mentioned in Section 3 or the results of [5], which show that the multiverse axioms of [9] are true amongst the countable computably-saturated models of set theory, in turn supports philosophical conclusions about the nature of the full set-theoretic multiverse.

The principle that pervasive features of the toy multiverses are evidence for the truth of those features in the full multiverse is a reflection principle similar in kind to those that are often used to provide philosophical justification for large cardinals. Just as those reflection principles regard the full universe V as fundamentally inaccessible, yet reflected in various much smaller pieces of the universe, the principle here regards the full multiverse as fundamentally inaccessible, yet appearing in part locally as a toy multiverse within a given universe. So our knowledge of what happens in the toy multiverses becomes evidence of what the full multiverse may be like.

Ultimately, the multiverse vision entails an upwardly extensible concept of set, where any current set-theoretic universe may be extended to a much larger, taller universe. The current universe becomes a countable model inside a larger universe, which has still larger extensions, some with large cardinals, some without, some with the continuum hypothesis, some without, some with $V = L$ and some without, in a series of further extensions continuing longer than we can imagine. Models that seem to have $0^\sharp$ are extended to larger models where that version of $0^\sharp$ no longer works as $0^\sharp$, in light of the new ordinals. Any given set-theoretic situation is seen as fundamentally compatible with $V = L$, if one is willing to make the move to a better, taller universe. Every set, every universe of sets, becomes both countable and constructible, if we wait long enough. Thus, the constructible universe L becomes a *rewarder of the patient*, revealing hidden constructibility structure for any given mathematical object or universe, if one should only extend the ordinals far enough beyond one's current set-theoretic universe. This perspective turns the $V \neq L$ via maximize argument on its head, for by maximizing the ordinals, we seem able to recover $V = L$ as often as we like, extending our current universe to larger and taller universes in diverse ways, attaining $V = L$ and destroying it in an on-again, off-again pattern, upward densely in the set-theoretic multiverse, as the ordinals build eternally upward, eventually exceeding any particular conception of them.

References

1. Jon Barwise. Infinitary logic and admissible sets. *J. Symbolic Logic*, 34(2):226–252, 1969.
2. Jon Barwise. Infinitary methods in the model theory of set theory. In *Logic Colloquium '69 (Proc. Summer School and Colloq., Manchester, 1969)*, pages 53–66. North-Holland, Amsterdam, 1971.
3. Jon Barwise. *Admissible sets and structures*. Springer-Verlag, Berlin, 1975. An approach to definability theory, Perspectives in Mathematical Logic.
4. Sy-David Friedman. Internal consistency and the inner model hypothesis. *Bull. Symbolic Logic*, 12(4):591–600, 2006.
5. Victoria Gitman and Joel David Hamkins. A natural model of the multiverse axioms. *Notre Dame J. Form. Log.*, 51(4):475–484, 2010.
6. Joel David Hamkins. Every countable model of set theory embeds into its own constructible universe. *to appear in Journal of Mathematical Logic*.
7. Joel David Hamkins. Is the dream solution of the continuum hypothesis attainable? pages 1–10. under review.
8. Joel David Hamkins. The set-theoretic multiverse : A natural context for set theory. *Annals of the Japan Association for Philosophy of Science*, 19:37–55, 2011.
9. Joel David Hamkins. The set-theoretic multiverse. *Review of Symbolic Logic*, 5:416–449, 2012.
10. Joel David Hamkins, David Linetsky, and Jonas Reitz. Pointwise definable models of set theory. *Journal of Symbolic Logic*, 78(1):139–156, 2013.
11. Daniel Isaacson. The reality of mathematics and the case of set theory. In Zsolt Novak and Andras Simonyi, editors, *Truth, Reference, and Realism*. Central European University Press, 2008.
12. Benedikt Löwe. A first glance at non-restrictiveness. *Philosophia Mathematica*, 9(3):347–354, 2001.
13. Benedikt Löwe. A second glance at non-restrictiveness. *Philosophia Mathematica*, 11(3):323–331, 2003.
14. Penelope Maddy. Believing the axioms, I. *The Journal of Symbolic Logic*, 53(2):481–511, 1988.
15. Penelope Maddy. Believing the axioms, II. *The Journal of Symbolic Logic*, 53(3):736–764, 1988.
16. Penelope Maddy. *Naturalism in Mathematics*. Oxford University Press, 1997.
17. Penelope Maddy. $V = L$ and MAXIMIZE. In *Logic Colloquium '95 (Haifa)*, volume 11 of *Lecture Notes Logic*, pages 134–152. Springer, Berlin, 1998.
18. Donald A. Martin. Multiple universes of sets and indeterminate truth values. *Topoi*, 20(1):5–16, 2001.
19. W. N. Reinhardt. Remarks on reflection principles, large cardinals, and elementary embeddings. *Proceedings of Symposia in Pure Mathematics*, 13(II):189–205, 1974.

20. Saharon Shelah. Logical dreams. *Bulletin of the American Mathematical Society*, 40:203–228, 2003.

21. John R. Steel. Generic absoluteness and the continuum problem. Slides for a talk at the Laguna workshop on philosophy and the continuum problem (P. Maddy and D. Malament organizers) March, 2004. http://math.berkeley.edu/~ steel/talks/laguna.ps.

HILBERT, BOURBAKI AND THE SCORNING OF LOGIC

A. R. D. Mathias

ERMIT, Université de la Réunion
UFR Sciences et Technologies
Laboratoire d'Informatique et de Mathématiques
2 rue Joseph Wetzel Bâtiment 2
F 97490 Sainte-Clotilde, France
ardm@univ-reunion.fr, ardm@dpmms.cam.ac.uk

In memoriam
Brian Wormald et Maurice Cowling,
Domus Divi Petri apud Cantabrigienses sociorum,
auctoris olim collegarum amicorumque,
virorum et humanitate et doctrina praestantium,
hoc opus grato animo dedicat auctor.

THE ARGUMENT

Those who wish to teach in the *lycées* and *collèges* of France must obtain a *Certificat d'Aptitude au Professorat de l'Enseignement du Second Degré*, commonly called the CAPES. For that certificate, the first hurdle to be passed is a written examination. In at least ten recent years the syllabus for the mathematics section of that examination, as specified in the relevant special numbers of the *Bulletin Officiel*, has contained the line

> Tout exposé de logique formelle est exclu.

That remarkable ban, still in force though latterly softened to *"Aucun exposé de logique formelle n'est envisagé"*, is a sign in the context of teacher-training of a more widely spread phenomenon. My main point will be pedagogical, that the teaching of logic, in France and elsewhere, at all academic levels, has over several decades been hampered, not to say crippled, by policies stemming from the lack, among members of the Bourbaki group,

47

of interest in and understanding of the foundations of mathematics; to put it starkly,

1. widely-read texts contain falsehoods and misconceptions about logic;
2. the prestige of their authors means that these errors are meekly accepted;
3. correct teaching of logic is thereby blocked.

Bourbaki in many things took Hilbert as their model, and it appears that their difficulties with logic are rooted in Hilbert's pre-Gödelian misconception of the relationship of truth to consistency. The story that I shall tell is this:

A: Hilbert in 1922, in joint work with Bernays, proposed an alternative treatment of predicate logic *49*

B: ... which, despite its many unsatisfactory aspects, was adopted by Bourbaki for their series of books *63*

C: ... and by Godement for his treatise on algebra, though leading him to express distrust of logic. *72*

D: It is this distrust, intensified to a phobia by the vehemence of Dieudonné's writings, ... *88*

E: ... and fostered by, for example, the errors and obscurities of a well-known undergraduate textbook, ... *98*

F: ... that has, I suggest, led to the exclusion of logic from the CAPES examination. *116*

G: Centralist rigidity has preserved the underlying confusion and consequently flawed teaching; *121*

H: ... the recovery will start when mathematicians adopt a post-Gödelian treatment of logic. *127*

Sections C, D, and E may be regarded as case studies, containing detailed criticisms of the logic portions of two algebra textbooks of the Bourbaki school and of various pronouncements on logic by Bourbaki's leading spokesman. The reader who wishes to defer reading such details should perhaps first read Sections A and B and then pass directly to Sections F, G, and H.

A: *Hilbert in 1922 proposes an alternative treatment of predicate logic ...*

THE CURTAIN RISES in Göttingen on 29·xii·1899,[*] to reveal Hilbert[1] writing to Frege[2] that

> if the arbitrarily given axioms do not contradict each other with all their consequences, then they are true and the things defined by the axioms exist. This is for me the criterion of truth and existence.[1]

A·1 It is hard to judge at this distance what Hilbert would have made of the many independence results found by set theorists in the later twentieth century. Though he knew very well that there are mutually inconsistent systems of geometry, he seems not to have considered that there might be two mutually contradictory systems of set theory, or even of arithmetic, each in itself consistent, so that the objects defined by the two sets of axioms cannot co-exist in the same mathematical universe.

Let us give some examples from set theory. Suppose we accept the system ZFC. Consider the following pairs of existential statements that might be added to it.

[A_1] the real number $0^\sharp$ exists.

[B_1] there are two Σ_1^1 sets of reals, neither of them a Borel set, and neither reducible to the other by a Borel isomorphism.

[A_2] there is a measurable cardinal.

[B_2] there is an undetermined analytic game.

[A_3] there is a supercompact cardinal.

[B_3] there is a projective well ordering of the continuum.

---

[*] To place the events of 1922, and their consequences for our enquiry, in context, we must mention some of the milestones in the development of logic and set theory in the early twentieth century. A complete account would be impossible, as each of the giants involved stood on the shoulders of other, earlier, giants; I ask the reader, and the historians of science whom I have consulted, to forgive the crudities of my perforce simplified narrative.

[1] David Hilbert, 1862–1943; Ph.D. Königsberg 1885; there till his move to Göttingen in 1895.

[2] Gottlob Frege, 1848–1925; Ph.D. Göttingen 1873; in Jena from 1874. According to [OH], a reluctant *conferencier* but an indefatigable correspondent.

[1] I quote the translation given in Kennedy [Ke], who there coins the apt phrase "Hilbert's Principle".

For each i, the statement A_i refutes the statement B_i in ZFC; but so far as is known each of the statements may consistently be added to ZFC, though some of them are stronger than others.

So Hilbert's Principle, that consistency is a ground for existence, should be taken to mean that a consistent theory describes *something*.

A·2 In 1900, at the Paris International Congress of Mathematicians, Hilbert said

> "This conviction of the solubility of every mathematical
> problem is a powerful incentive to the worker. We hear
> within us the perpetual call: *There is the problem. Seek
> its solution. You can find it by pure reason, for in math-
> ematics there is no ignorabimus.*"

A·3 On 16·vi·1902, Russell wrote to Frege to communicate his discovery, in 1901, of a contradiction in Frege's theory of classes; Frege acknowledged the contradiction in the second volume of his *Grundgesetze der Arithmetik*, published in 1903; on 7·xi·1903 Hilbert wrote to Frege to say that such paradoxes were already known in Göttingen, discovered by Hilbert and by Zermelo[3],[2] and that

> "they led me to the conviction that traditional logic is in-
> adequate and that the theory of concept-formation needs
> to be sharpened and refined."

A·4 On 18·ix·1904, Zermelo wrote to Hilbert to communicate his first proof that every set can be well ordered; his letter was published as [Ze1]. In response to criticisms of his proof, Zermelo in 1908 published in [Ze2] a re-working of it and in [Ze3] a proposal for a system of axioms for basing mathematics on set theory, though it should be remembered that as the process of formalising such systems was still in its infancy, Zermelo had to leave undefined the concept of *definite Eigenschaft* invoked in his Separation scheme.

A·5 In lectures at Göttingen in 1905, Hilbert, after discussing set theory and the paradoxes, said[3]

> "The paradoxes we have just introduced show sufficiently
> that an examination and redevelopment of the founda-
> tions of mathematics and logic is urgently necessary."

---

[3] Ernst Zermelo, 1871–1953; Ph.D. Berlin 1894; in Göttingen 1897–1910, Zürich 1910–1916, (retirement, with a Cantonal pension, forced by ill-health); honorary Professor, Freiburg 1926–35 and 1946–53.

[2] See [RaTh] and, for a more recent discussion, [Pec].

[3] as translated in [Za1], page 333.

A·6 But from 1905 for twelve years or so Hilbert was absorbed in other projects, such as his work on the axiomatization of physics, described in [Cor4], and his work on integral equations. To illustrate the breadth of his interests we quote three trenchant remarks from various periods of his life.

On the eternally shifting balance between geometry and arithmetic: Weierstrass thought all reduced to number;[4] but Hilbert countered by saying in his lectures, in Königsberg in 1891, that "Geometry deals with the properties of space, [perception of which comes to us] through the senses."[5]

Hilbert's remark, in lecture notes of 1894,[6] that "Geometry is a science whose essentials are developed to such a degree that all its facts can already be logically deduced from earlier ones. Much different is the case with the theory of electricity" suggests that to him axiomatisation is the last stage of a process which begins with the amassing of data and ideas.

Hilbert's enthusiasm for physics is evident: "Newtonian attraction turned into a property of the world-geometry, and the Pythagorean theorem into a special approximated consequence of a physical law."[7]

1917: Hilbert returns to the foundations of mathematics

A·7 Fast forward now to Göttingen in the Winter Semester of 1917, to discover Hilbert giving a course of lectures on the foundations of mathematics.

That is in itself noteworthy. Many modern mathematicians of distinction would decline to devote time and effort to the foundations of their subject; but plainly Hilbert would not waste his time on things he judged unimportant. In his earlier days he had been thrilled to adopt the set-theoretical ideas of Cantor as a framework for mathematics; and his much-quoted remark of 1926 that no one shall drive us from the paradise that Cantor created, though directed against the intuitionist and constructivist reactions of Brouwer[4] and Weyl[5], of 1910 and 1918 respectively, may perhaps be seen as an affirmation that the foundations of classical mathematics have been re-built and rendered impregnable after the set-back of the discovery of paradoxes in Frege's detailed treatment of the theory of classes.

---

[4] [Cor5], p 168.

[5] [Cor5], p 156-7.

[6] translated in [Cor3], page 257.

[7] In *Wissen und mathematisches Denken*, 1922-3, translated in [Cor5], p 173.

[4] Luitzen Egbertus Jan Brouwer, 1881–1966; Ph.D. 1907, Amsterdam; extr. professor, Amsterdam 1912–1951.

[5] Hermann Klaus Hugo Weyl, 1885–1955; Ph.D. Göttingen, 1908; there till 1913 and 1930–33; Professor, ETH, Zürich 1913–1930; IAS Princeton, 1933-1952.

As Hilbert's course progressed, typewritten notes of his lectures were prepared with scrupulous care by his assistant, Bernays[6]; these notes are preserved in the archives of Göttingen, and, some years later, were used by Ackermann,[7] a pupil of Hilbert, in preparing the text [HiA], to which we return below, for its publication in 1928. Their continued influence can perhaps be detected in the first volume of the much larger, two-volume, treatise [HiB1,2], discussed towards the end of this section.

The notes of 1917/18 then gradually disappeared from view, until in recent years Sieg and his collaborators Ewald, Hallett and Majer have initiated and nearly completed their re-appraisal.[8]

The reader of these lecture notes[9] will notice, underneath the period style, the modernity of the conception of logic that is being expounded, though of course many results, such as the completeness theorem for predicate logic, had not yet been proved and still had the status of open questions; and other results, particularly the incompleteness theorems for systems of mathematics, were undreamt of.

Some terminology

Indeed, the notion of a formal system of mathematics had made great strides since 1905, and it will be helpful, without going into either detail or history, to review some of the vocabulary of modern logic. We suppose that we have already specified an appropriate formal language (which means specifying the symbols of the language and the rules of formation of its formulæ), and that we have specified the underlying logic and rules of inference. A *sentence* of such a language is a formula without free variables. A *theory* in such a language will then be specified by choosing the sentences that are to be its axioms.

Four possible properties of such a theory are consistency, syntactic completeness, semantic completeness and decidability.

A·8 DEFINITION A theory is *consistent* if no contradiction can be derived in it; a consistent theory is said to be *syntactically complete* if no further

---

[6] Paul Bernays, 1888–1977; Ph.D. Berlin 1912; Zürich 1912–1917 (Habil. 1912; assistant to Zermelo) and from 1934; Göttingen 1917–1933, (Habil. 1918, Extraordinary Professor from 1922).

[7] Wilhelm Ackermann, 1896–1962; Ph.D. Göttingen 1925; from 1929 schoolmaster at Burgsteinfurt & Lüdenscheid.

[8] A critical edition [ES] of these notes is being prepared by Ewald and Sieg for publication; pending their publication, Sieg's paper [Si1] may be consulted for much historical and mathematical detail.

[9] I am grateful to Professor Sieg for placing a copy at my disposal.

axioms can be added without an inconsistency resulting. Thus in a syntactically complete theory every sentence of its language is either provable or refutable.

The above properties of a theory can be understood knowing only its axioms and rules of inference without knowing anything about its intended interpretations.

The semantic counterparts to those properties are defined in terms of the intended interpretations of the theory in question, which must therefore be specified; often it is enough to consider interpretations in finite non-empty domains and in a countably infinite domain.

A·9 DEFINITION A sentence of the theory is *universally valid* if true in all its intended interpretations. A theory is *sound* if all its theorems are universally valid. The theory is *semantically complete* if its every universally valid sentence is a theorem. Put another way, if a sentence is irrefutable it is true in at least one of the theory's intended interpretations.

The above are thus properties of the theory relative to its specified family of intended interpretations.

The *completeness theorem* of Gödel, proved in his dissertation of 1929 and published in 1930, says that every consistent first-order theory in a countable language has a countably infinite model (or possibly a finite one, if one has undertaken to interpret the equality predicate = as identity.)

Thus if all such models are counted as intended interpretations, the theory is semantically complete; and the completeness theorem for such theories yields Hilbert's Principle.

A·10 EXAMPLE The theory of non-empty endless dense linear orderings. There are no finite models of this theory, and Cantor proved that any denumerable such must be isomorphic to ℚ, the set of rational numbers with its usual ordering; so the theorems of this theory are precisely the sentences in the language of linear orderings true in ℚ; as every sentence is either true or false in that model, the theory is syntactically complete.

A·11 REMARK The reader should be warned that "complete" was often used by Hilbert to mean "covers everything known so far"; so a theory thought to be complete in his sense today might be seen as incomplete tomorrow.

A·12 DEFINITION A theory is *decidable* if there is an algorithm which given any sentence of the theory will decide in finite time whether or not it is a theorem. For classical propositional logic, such an algorithm exists; but that is not true for classical first-order predicate logic.

1928: publication of the treatise of Hilbert and Ackermann

We mention this text here as it belongs to a stage in the development of logic rather earlier than its date of publication: Hilbert in his foreword, dated 16·i·1928, states that the sources used are the 1917/18 notes, together with notes on courses given in the winter semesters of 1920 and 1921/2. The delay in its publication may perhaps be attributed to Hilbert's absorption in the ε-operator that he defined in 1922: for he remarks that the book should serve as preparation and lightening of a further book that he and Bernays wish to publish soon which will treat the foundations of mathematics using the epsilon symbol.

The contents of the book are well summarised in [Bu]; there is a thorough treatment of propositional logic in Chapter One, using essentially the axioms given in *Principia Mathematica*; simplifications due to Bernays are used in the axiomatisation of predicate logic given in Chapter III, and its consistency and syntactic incompleteness proved; but the rest of the book largely consists of examples, as the semantic completeness of predicate logic had not yet been proved. Though that form of completeness is indeed mentioned as an open problem, Hilbert's Principle would seem to have been an article of faith: Hilbert thought in 1919 that "things cannot be otherwise";[10] true of a complete theory, but not of an incomplete one, which might have more than one completion.

Logic in the twenties

The early years of the twentieth century were a time of intense activity in foundational research, and a sense of the variety of proposals for predicate logic may be obtained from Goldfarb's 1979 paper [Gol].[11]

Besides those of Hilbert, Goldfarb discusses the accounts of predicate logic offered by Frege (1882, 1892), Russell (1903, 1919), Schröder (1895), Löwenheim (1915), Skolem[8] (1920), (1922), Herbrand[9] (1928, 1930), and Gödel[10] (1930). He sees the twenties as a period in which the ideas of two schools of logic originating in the nineteenth century:

---

[10] [Cor5], page 156.

[11] Goldfarb was writing without knowledge of the 1917/18 notes of Hilbert, and therefore, as shown by Sieg in [Si 1], his chronology in places requires correction.

[8] Thoralf Skolem, 1887–1963; Göttingen 1915; Oslo 1916–1930 (late Ph.D. 1926) and from 1938; Bergen 1930–38.

[9] Jacques Herbrand, 1908–1931; Ph.D. Sorbonne, 1930; visited Berlin, Hamburg, Göttingen 1931; killed climbing.

[10] Kurt Gödel, 1906–1978; in Vienna from '24 (Ph.D. '30, Habil. '32); visited US '33, '35; IAS, Princeton from '39.

1) the algebraists : de Morgan (1806–71), Boole (1815–64), Peirce (1839–1914), Schröder (1841–1902), Löwenheim (1878–1957);

2) the logicists : Frege, Peano (1858–1932), Russell (1872–1970).

merged to yield the modern theory of quantification. He writes:

> *The deficiencies in the two early traditions I have been discussing may be summarized thus. To arrive at meta-mathematics from Russell's approach we must add the "meta", that is, the possibility of examining logical systems from an external stand point. To arrive at meta-mathematics from the algebra of logic we must add the "mathematics", that is, an accurate appreciation of how the system may be used to encode mathematics, and hence of how our metasystematic analyses can be taken to be about mathematics.*

One might add that the idea, so well-established today, that one might wish to interpret a formula in many different structures, came more easily to the algebraists than to the logicists. The two schools worked in partial knowledge of each other's efforts;[12] the discovery of the *quantifier* is generally attributed to Frege (1879) but was made independently and slightly later ([Mi], 1883) by O. H. Mitchell, a student of Peirce, who further developed the idea in a paper [Pei] of 1885.[13]

1922: the Hilbert operator is launched

We turn to the proposal made by Hilbert in 1922; not because of any alleged superiority to the other contemporary accounts of logic but because that was the one adopted by Bourbaki; and indeed the conclusion to which we shall come is that Bourbaki backed the wrong horse.

Hilbert's 1922 proposal was based on what is often called the Hilbert ε-operator, by which quantifiers could formally be avoided and predicate logic reduced to propositional.** The apparent simplicity of this proposal commended it to the members of Bourbaki, who made it, though in a different notation, the basis of their Volume One, on the theory of sets, and developed it as the logical basis of their series of books. We review its history now and defer to Section B a discussion of its demerits.

---

[12] For a portrait of the mutual non-admiration of Peirce and Russell, see the paper [An].

[13] A helpful introduction to the early history of the quantifier is the paper [Pu].

** But *something* is being concealed, as propositional logic is decidable whereas predicate logic is not.

τ versus ε:[14]

A·13 In a lecture [Hi1][15] given in September 1922, Hilbert, describing joint work with Bernays, introduced what he called a logical function which he wrote as $\tau(\mathfrak{A})$ or $\tau_a(\mathfrak{A}(a))$, which associates to each one-place predicate $\mathfrak{A}(a)$ an object $\tau(\mathfrak{A})$. He gave its intended meaning in his Axiom 11, (which he credits to Bernays)

$$\mathfrak{A}(\tau\mathfrak{A}) \implies \mathfrak{A}(a)$$

and illustrated its use by saying that if $\mathfrak{A}$ were the predicate "is bribable", then $\tau\mathfrak{A}$ would be understood to denote a man of such unassailable uprightness that were *he* bribable then must all mankind be too.

Hilbert then defines the quantifiers in terms of his operator.

Using his operator Hilbert went on to sketch a proof of the consistency of a weak version of arithmetic with a single function symbol ϕ defined by recursion equations not involving the symbol τ. His idea was, roughly, a priority argument: *start by assigning* 0 *as the value to all τ-terms; redefine whenever a contradiction is reached; end by showing that you cannot have* $0 \neq 0$.

Hilbert plainly intended his operator to be the lynchpin of the new proof theory that he and his collaborator Bernays had set themselves to develop; he apologises for lacking the space in which to give all the details, but is evidently confident that their new theory will be able to dispel all the recent doubts about the certainty of mathematics.

A·14 In his inaugural dissertation, [Ack1][16], Ackermann reworked Hilbert's proof with greater care, found he needed to restrict the system yet further for the proof to work, and worked with the dual operator, which supplies a witness to an existential statement rather than a counter-example to a universal one; he named that operator ε, not τ. Thus his corresponding axiom reads

$$\mathfrak{A}(a) \implies \mathfrak{A}(\varepsilon_a\mathfrak{A}(a)).$$

This change of letter and operator was thenceforth adopted, except by Bourbaki who followed the change of operator without changing the letter.

---

[14] Much of this subsection has been gleaned from van Heijenoort's anthology [vHe]. For vastly improved detail, see [Zac2].

[15] manuscript received by *Mathematische Annalen* on 29.ix.1922 and published the following year.

[16] manuscript received 30.iii.1924 and published that year.

In [vN],[17] von Neumann[11] criticised Ackermann's paper and gave a consistency proof for first-order number theory with induction for quantifier-free formulæ. Hilbert in 1927 gave a lecture [Hi3], outlining Ackermann's paper and the method of "assign values then change your mind".

1928: Hilbert at Bologna

In his address [Hi4] on 3·ix·1928, to the International Congress of Mathematicians at Bologna,[♣] Hilbert reiterated his belief in the consistency, completeness and decidability of mathematics. He remarks that as mathematics is needed as the foundation of all the sciences, it is incumbent on mathematicians to secure its foundations. He hints at Skolem and Fraenkel having completed the axiomatisation of Zermelo; he mentions the ε-axiom, in Ackermann's notation, and the work of Ackermann and von Neumann's work on ε, which he seems to think has established the consistency of arithmetic; and he discusses four problems.

The first is to extend the proof of consistency of his ε-axioms to a wider class of formulæ; the second is to establish the consistency of a global, extensional, form of choice as expressed through his symbol; the third is to establish the completeness of axiom systems for arithmetic and for analysis; and the fourth is to establish the completeness of predicate logic: is everything that is always true a theorem?

In closing he remarks that we need mathematics to be absolutely true, otherwise *Okkultismus* might result; and he repeats his belief that *in der Mathematik gibt es kein Ignorabimus.*

1928: the war of the Frogs and the Mice

Brouwer gave two lectures in Vienna early in 1928 which according to an entry in Carnap's diary stimulated the young Gödel. Though Hilbert had helped Brouwer in his early career, their relationship had soured; personal tensions mounted in 1928, when Hilbert campaigned for and Brouwer against German participation in the Bologna Congress[18]; and matters came to a head in October 1928.[19] Even though Einstein chaffed Hilbert with the

---

[17] manuscript received 29.vii.1925 and published in 1927.

[11] Johann von Neumann, (1903-1957); Göttingen 1926/7 then Berlin; Hamburg 1929/30, thereafter in Princeton.

[♣] with Henri Cartan present, but not Gödel, who spent that summer in Brno reading *Principia Mathematica.*

[18] [Seg2], pp 352, 354.

[19] [vDa1; vDa2, Volume II].

phrase *Froschmaüsekrieg*, in the hope of restoring calm, the outcome was the expulsion of Brouwer, after thirteen years' service, from the Editorial Board of the *Mathematische Annalen*.

Brouwer retaliated by launching the journal *Compositio Mathematica*.

1929: the completeness theorem

The fourth of Hilbert's Bologna problems was solved affirmatively the following year by Gödel in his doctoral dissertation.[20] It would seem that Skolem had earlier come close to a proof, and that to some extent his modesty has obscured the chronicle:♠ Gödel in 1964 wrote that Skolem in his 1922 paper [Sk] proved (but did not clearly state) the result that if a formula is not a theorem its negation is satisfiable. Syntactic completeness might be finitistic: one could imagine an algorithm which given a non-theorem finds a proof of its negation. But semantic completeness is not: as the models in which formulæ are to be tested are countable but perhaps not finite, an infinite sequence of admittedly finitistic steps is needed to build one. Gödel thought that Skolem had the steps but not the sequence. Goldfarb [Gol] writes:

> "*Gödel's doctoral dissertation and its shorter published version ... is a fitting conclusion to the logic of the twenties. [...] Although Gödel [worked] independently, the mathematics is not new: it was substantially present in the work of both Skolem and Herbrand. What is new is the absolute clarity Gödel brings to the discussion.*"

For a more recent study that also conveys very clearly the impact of Gödel's dissertation, see [Ke].

1930/31: the incompleteness theorems

The third of Hilbert's Bologna problems was solved negatively by the two incompleteness theorems of Gödel. It is tempting to speculate that Hilbert was initially misled by his success in giving a formal treatment of Euclidean plane geometry into thinking that the corresponding foundational problem posed by arithmetic would prove to be similar, so that the consistency and completeness of mathematics could be established by finitistic means. Goldfarb [Gol] highlights that misconception:

---

[20] All cited papers and correspondence of Gödel will be found in the Collected Works [Gö].

♠ as has Bernays' modesty, as shown by Zach [Za1], the history of propositional logic.

"By the end of the decade the Hilbert school was quite certain that they had in all essentials a [consistency] proof for full number theory. Gödel's Second Incompleteness Theorem came as a terrible shock."

1931/34: Hilbert's delayed response to the incompleteness theorems

None perhaps was more shocked than Hilbert. A recent paper of Sieg [Si4] documents the strange contrast between the admirably enthusiastic and generous response of von Neumann to Gödel's results and the rather less admirable state of denial that was Hilbert's initial response.

Hilbert and Gödel were at two different meetings in Königsberg in East Prussia in September 1930. On September 7th, at a round-table discussion, moderated by Hans Hahn and attended by von Neumann and Carnap among others, which discussion formed part of the second *Tagung für Erkenntnislehre der exakten Wissenschaften*, Gödel announced his first incompleteness theorem. Gödel and von Neumann discussed this result immediately after the session, and then by mid-November each of them had independently found the second incompleteness theorem.

On September 8th, Hilbert gave his famous lecture *Naturerkennen und Logik* to a meeting of the *Gesellschaft Deutscher Naturforscher und Ärzte*.

Gödel left Königsberg on September 9th, without, it seems, having discussed his new result with Hilbert.[21]

I follow Sieg in thinking that it is implausible that von Neumann should not have spoken to Hilbert about it, for given that Hilbert had said at Bologna that von Neumann and Ackermann had a proof of the consistency of arithmetic, then undoubtedly von Neumann, once its impossibility was clear to him, would immediately have disabused him of this idea.

The strange thing is that Hilbert published two papers [Hi5] and [Hi6], on foundational themes after Gödel's announcement without explicitly mentioning Gödel's work. Both were published in 1931; [Hi5] is the text of a lecture given to the *Philosophischen Gesellschaft* of Hamburg in December 1930 and was received by the editorial board of *Mathematische Annalen* on December 21st, 1930; whereas [Hi6] was presented to the Göttingen *Gesellschaft der Wissenschaften* on July 17th, 1931.

In [Hi5] from page 492 on, Hilbert gets defensive, suggesting an awareness of Gödel's results, but writes defiantly on page 494:

---

[21] In a letter cited in footnote 4 of [Daw], Gödel states that he neither met nor ever corresponded with Hilbert.

> *"Ich glaube, das, was ich wollte and versprach, durch die*
> *Beweistheorie vollständig erreicht zu haben: Die mathe-*
> *matische Grundlagenfrage als solche ist dadurch, wie ich*
> *glaube, endgültig aus der Welt geschafft."* [i]

In [Hi6], on page 122, he writes:

> *"Nunmehr behaupte ich, daß "widerspruchsfrei" mit "richtig"*
> *identisch ist."* [ii]

When Bernays in an interview given on August 17th, 1977, three weeks before his death, cited in [Si4], was asked *"Wie hat Hilbert reagiert, als er von Gödels Beweis der Unmöglichkeit, einen Widerspruchsfreiheitsbeweis für die Zahlentheorie im Rahmen der Zahlentheorie selbst zu führen, erfuhr?"* [iii], he replied *"Ja, ja, er war ziemlich ärgerlich darüber ... Aber er hat nicht bloss negativ reagiert, sondern er hat ja dann eben auch Erweiterungen vorgenommen, z.B. schon im Hamburger Vortrag von 1930"* [iv].

Hilbert's programme after Gödel

It took some years for Gödel's 1931 paper to be generally taken on board[22]: although von Neumann and Herbrand grasped the point quickly, Zermelo did not. One might ask, "After this check, what is left of Hilbert's programme?"

The first reaction of many, including Herbrand,[23] was to think that the incompleteness theorem showed the impossibility of Hilbert's programme. The view taken by Bernays and Gödel was that suggested by Hilbert in his foreword to [HiB1]: there might be finitist arguments not formalisable in Peano arithmetic.

Goldfarb writes that Herbrand's papers (1929, 1930) made an important contribution to Hilbert's programme which led to the Hilbert–Bernays ε theorems, even though one of Herbrand's arguments is fallacious; further,

---

[i] "I believe that through proof theory I have completely achieved what I wanted and promised: foundational questions about mathematics are, so I believe, finally expelled from the world."

[ii] "Further, I assert that "consistent" is identical with "true".

[iii] "How did Hilbert react when he realised that Gödel's proof showed that a consistency proof for number theory could not be given within number theory?"

[iv] "Yes, yes, he was fairly cross about it .... but his reactions weren't only negative: already in his lecture in Hamburg in 1930, he developed extensions of it."

[22] See [Daw].

[23] Sieg in [Si2] examines the brief exchange of letters in 1931 between Herbrand and Gödel.

Herbrand's work was largely the foundation for Gentzen's Hauptsatz, which substantiated Hilbert's hope that finitism would go further.[24] Ackermann in [Ack2] showed that Gentzen's 1936 proof [G2] of the consistency of arithmetic by an induction up to the ordinal ε_0 could be presented in terms of the Hilbert-Bernays-Ackermann operator.[25]

Sieg in [Si4] describes not only the discomfiture of Hilbert but on the positive side the advent of the young Gentzen and the further growth of proof theory in a new direction beyond Hilbert's original conception.

Thus Hilbert's programme [Si1] was not wasted but developed into proof theory [Si3].

1934, 1939: publication in two volumes of the treatise of Hilbert and Bernays

Both these volumes were, in fact, written entirely by Bernays, by 1934 expelled from Göttingen by the newly-elected Nazi Government of Germany, and residing once more in Zürich.

A·15 The first volume [HiB1], of 1934, of the treatise of Hilbert and Bernays, which may be regarded as an expanded version of [HiA], presents a treatment of first-order logic without the operator, and includes a mature account of the theory of recursion.

Hilbert in his foreword maintains that the view that the results of Gödel entail the impossibility of Hilbert's programme of proof theory is erroneous: they merely make necessary a more precise account of finitism. Bernays in his explains that it had become necessary to divide the projected book mentioned in the foreword to [HiA] into two parts, partly as a result of the works of Herbrand and of Gödel. Both forewords are dated March 1934, with no exact day given.

A·16 The foreword of Paul Bernays to [HiB2] is dated February 1939. This second volume gives a detailed development of the epsilon calculus and proof of the two epsilon theorems; then proofs of Gödel's two incompleteness theorems. The very brief foreword of Hilbert to that second volume is dated March 1939 and makes no mention of Gödel.

In §2, 4, f), starting on page 121, Bernays presents an example due to von Neumann which pinpoints the error in the earlier alleged consistency proof for arithmetic.

---

[24] Black's review [Bl1] of both [HiB2] and a 1938 survey paper [G3] of Gentzen explains this point very clearly.

[25] For a later study of Hilbert's operator, see Leisenring [L], and for further historical perspective, see [DrKa].

1935: the naissance of Bourbaki

It was at this delicate period in the history of mathematical logic, when the subject was realigning itself following Gödel's discoveries, that the Bourbaki group was formed. There are many accounts in print of the group's inception, character and achievements,[26] so it is not necessary to repeat that saga here.♡ Our concern is with the group's chosen treatment of logic.

Hilbert's account of logic had received a considerable set-back. He had based his strategy on the belief that all problems could be solved within a single framework. There were substantial texts expounding his ideas; but the hope of a single proof of the consistency and completeness of mathematics, in my view the only justification for basing an encyclopædic account of mathematics on Hilbert's operator, had been dashed.

Constance Reid in Chapter XXI of her life [Re1] of Hilbert discusses the pernicious anæmia diagnosed in him in late 1925. A diet of raw liver seems to have saved him from the worst consequences, but the disease might well have sapped his strength, and more and more he relied on his younger colleagues to carry out his research programmes. When the incompleteness theorems appeared, he did respond to them in time, but, it would seem, only with reluctance.

Undeterred, and unfortunately, Bourbaki, as we shall see, adopted Hilbert's pre-Gödelian stance. In the next section we shall examine Bourbaki's account as finally presented; and in Section H explore the soul-searching, revealed in the recently available archives of Bourbaki, among members of the Bourbaki group that led to that final chosen position.

---

[26] such as [Bea1], [Bea2], [Cor1], [Cor2], [Cor3], [Mas1], [Mas2], [Bor], [Cho], [Pl8], and [Sen].

♡ We record a recent suggestion concerning the choice of the group's pseudonym. The name Bourbaki(s), though a Cretan surname, is composed of (the French transliteration of) the two Hebrew words 'bour', meaning an ignoramus, and 'baki', meaning a wise man, so that certain members of the French group might have been enjoying a private joke in so modestly calling their collective self a sage dunce or an ignorant savant. The link with the general of Cretan descent in the service of Napoleon III might then have been a convenient cover story, very necessary in Nazi-occupied Paris.

B: ... which is adopted by Bourbaki ...

BOURBAKI'S SYNTAX, which we shall now outline, was followed by Godement, whose treatment of logic in his *Cours d'Algèbre* we examine in the next section; we follow their notation in our discussion. Citations such as E I.34 are from Bourbaki's text *Théorie des ensembles* [Bou54].[27]

There were position papers on the foundations of mathematics, published in 1939 and 1943, by members of the Bourbaki group; and on December 31st, 1948, in Columbus, Ohio, Nicolas Bourbaki, by invitation, addressed the eleventh meeting of the Association for Symbolic Logic. That address, [Bou49], chaired by Saunders Mac Lane, delivered by André Weil and published the following year, presents the system that is discussed in [M10] and there called Bou49. The book on logic and set theory that Bourbaki had, after an initial reluctance[28], by then decided to include as Livre I of their projected series, was published by chapters, in 1954, '56 and '57. We denote by Bou54 the system of set theory developed in [Bou54].

Some differences between the two: in 1949, Bourbaki made no mention of the Hilbert operator, and claimed to be able to base "all existing mathematics" on an axiomatic system broadly similar to that of Zermelo 1908, a claim modestly reduced to "modern analysis" in [Di1]. They take ordered pair as a primitive, and appear to believe that the existence of unordered pairs will then follow, a belief refuted in [M10]. They present their underlying system of logic by introducing the notion of a true formula and of the synonymy of two formulæ; they, in effect, state that all propositional tautologies are to be axioms; and they give axioms for quantifier logic, their treatment differing both from [HiA] and from Gödel's completeness paper.

By 1954 on the other hand, they had enhanced their axiom scheme of union to imply a form of the axiom scheme of replacement, they had refined their treatment of propositional logic,[∥] giving the same four axioms as those on page 22 of [HiA], which itself follows closely the treatment in *Principia Mathematica* [WR], where these four axioms are given, besides a fifth shown in 1918 by Bernays [Ber1] to be redundant:

---

[27] We shall occasionally refer to Bourbaki's in-house journal *La Tribu*, but defer comment on these and other revealing archives, now on-line, till Section H, as our immediate concern is with Bourbaki's books as published.

[28] The first formal gathering of Bourbaki, at Besse-en-Chandesse in 1935, resolved to give no axioms for set theory.

∥ possibly encouraged by the proof in [Hu] that the propositional logics of [WR] and [HiB1] coïncide.

$$p \lor p \centerdot \supset \centerdot p, \quad q \centerdot \supset \centerdot p \lor q, \quad p \lor q \centerdot \supset \centerdot q \lor p,$$
$$q \supset r \centerdot \supset \, \colon p \lor q \centerdot \supset \centerdot p \lor r$$

and, most significantly, they based their predicate logic no longer on quantifiers but on the Hilbert operator,$^\diamond$ a puzzling change as Hilbert himself, in his text with Ackermann and in the first volume of his text with Bernays, presented a development of predicate logic that is operator-free. According to the archives of Bourbaki, the Hilbert operation was missing from Draft 4 of Chapter I, but is found in Draft 5, presumably by Chevalley, of July 1950, and kept in Draft 6, by Dixmier, of March 1951.

Bourbaki's formal language admitted a potentially infinite supply of letters. In 1954 they kept ordered pair as a primitive, written $\supset$, but in later editions followed Kuratowski and defined $(x, y) = \{\{x\}, \{x, y\}\}$, where the unordered pair $\{x, y\}$ is the set whose sole members are x and y, and the singleton $\{x\}$ is $\{x, x\}$.

Bourbaki's syntax

B·1 Bourbaki use Ackermann's dual operator but write it for typographical reasons as τ rather than ε.

Bourbaki use the word *assemblage*, or, in their English translation, *assembly*, to mean a finite sequence of signs or letters, the signs being τ, $\square$, $\lor$, $\neg$, $=$, $\in$ and, in their first edition, $\supset$. The substitution of the assembly A for each occurrence of the letter x in the assembly B is denoted by $(A|x)B$.

Bourbaki use the word *relation* to mean what Anglophones would call a well-formed formula.

B·2 The rules of formation for τ-terms are these:

let R be an assembly and x a letter; then the assembly $\tau_x(R)$ is obtained in three steps:

(B·2·0) form τR, of length one more than that of R;

(B·2·1) link that first occurrence of τ to all occurrences of x in R

(B·2·2) replace all those occurrences of x by an occurrence of $\square$.

In the result x does not occur. The point of that is that there are no bound variables; as variables become bound (by an occurrence of τ,) they are replaced by $\square$, and those occurrences of $\square$ are linked to the occurrence of τ that binds them.

The intended meaning is that $\tau_x(R)$ is some x of which R is true.

$\diamond$ They would not have desired the heavily type-theoretic treatment of predicate logic in *Principia Mathematica*.

Certain assemblies are *terms* and certain are *relations*. These two classes are defined by a simultaneous recursion, presented in Godement [Gd] in nine clauses, thus:

T1: every letter is a term

T2: if A and B are terms, the assembly $\supset AB$, in practice written (A, B), is a term.

T3: if A and T are terms and x a letter, then $(A|x)T$ is a term.

T4: if R is a relation, and x a letter, then $\tau_x(R)$ is a term.

R1: If R and S are relations, the assembly $\vee RS$ is a relation; in practice it will be written $(R \vee S)$.

R2: $\neg R$ is a relation if R is.

R3: if R is a relation, x a letter, and A a term, then the assembly $(A|x)R$ is a relation.

R4: If A and B are terms, $=AB$ is a relation, in practice written $A = B$.

R5: If A and B are terms, the assembly $\in AB$ is a relation, in practice written $A \in B$.

B·3 REMARK Clauses T3 and R3 are, as pointed out to me by Solovay, redundant — if omitted, they can be established as theorems — and were added to Bourbaki's original definition by Godement, presumably for pedagogical reasons.

B·4 REMARK Note that every term begins with a letter, $\supset$ or τ; every relation begins with $=$, $\in$, $\vee$, or $\neg$. Hence no term is a relation.

Quantifiers are introduced as follows:

B·5 DEFINITION $(\exists x)R$ is $(\tau_x(R) \mid x)R$;

B·6 DEFINITION $(\forall x)R$ is $\neg(\exists x)\neg R$.

Thus in this formalism quantifiers are not primitive. Informally, the idea is to choose at the outset, for any formula $\Phi(x)$ a witness, some a such that $\Phi(a)$; call it $\tau_x\Phi$. If there is no such witness, let $\tau_x\Phi$ be anything you like, say the empty set.

B·7 We pause to consider some consequences of basing a formal system on the operator developed by Hilbert, Bernays and Ackermann, commonly called the Hilbert operator and used by Bourbaki.

The length of τ-expansions

Notice first that if we translate $\exists x\Phi$ as $\Phi(\tau_x\Phi)$, then in writing out that latter formula, we must replace every free occurrence of x in Φ by the string $\tau_x\Phi$, so that formulæ with many quantifiers become, when expanded, inordinately long. In [M8], it is shown that Bourbaki's definition of the number one, when expanded to their primitive notation, requires 4523659424929 symbols, together with 1179618517981 disambiguatory links.

To conceptualise the formalism becomes even more hopeless in later editions of Bourbaki, where the ordered pair (x, y) is introduced by Kuratowski's definition, not as a primitive, and the term for 1 takes an impressive 2409875496393137472149767527877436912979508338752092897 symbols,[29] with 871880233733949069946182804910912227472430953034182177 links.

Strangely, the number 1 takes much longer to define than the concept of the Cartesian product of two sets: for example in that formalism of Bourbaki's later editions, the term $X \times Y$ proves to be roughly of length 3.185×10^{18} with 1.151×10^{18} links, and 6.982×10^{14} occurrences each of X and Y; whereas the term defining 1 has over 2×10^{54} symbols with nearly 9×10^{53} links.

Every null term is equal to a proper term

Let us, following Leisenring's 1969 study [L], call a term $\tau_x(R)$ *null* if there is no x with the property R; and let us call it *proper* otherwise: we ignore the problem that there may be some terms of the status of which we know nothing. Is every term T, proper or otherwise, included in the range of variables? It is, and to see that, start from the

B·8 PROPOSITION $\forall a \exists b \, b = a$.

Now Criterion C30 on page E I.34 states that $(\forall x)R \implies (T \mid x)R$ is a theorem whenever T is a term, x a letter and R a relation. Hence we have, substituting T for a,

B·9 COROLLARY $\exists b \, b = T$.

Now let β be the term $\tau_b b = T$. β is a proper term, for we have proved there is such a b; and thus we have

B·10 COROLLARY $\beta = T$.

Any two null terms are equal

The principle S7, on page E I.38 of the *Théorie des Ensembles*, says that

> *si R et S sont des relations de $\mathcal{T}$ et x une lettre, la relation*
>
> $$((\forall x)(R \iff S)) \implies (\tau_x(R) = \tau_x(S))$$
>
> *est un axiome.*

They add, on page E I.39, "*le lecteur notera que la présence dans S7 du quantificateur $\forall x$ est essentielle.*"

---

[29] according to a program written by Solovay in Allegro Common Lisp.

B·11 REMARK A few lines lower, in small type, is the following remark:

> *Par abus de langage, lorsqu'on a démontré une relation*
> *de la forme $T = U$ dans une théorie $\mathcal{T}$, on dit souvent que*
> *les termes T et U sont "les mêmes" ou sont "identiques".*
> *De même, lorsque $T \neq U$ est vraie dans $\mathcal{T}$, on dit que T*
> *et U sont "distincts" au lieu de dire que T est différent*
> *de U.*

I wish that Bourbaki had not said that; I wish that they had reserved "identical" for the stricter meaning, and been happy with the phrase "$\mathcal{T}$-equivalent" for the looser meaning; because in a passage on page E II.4, which I quote below, they actually mean that two relations are identical, the very same formula, and not merely that they are provably equivalent.

B·12 We note a consequence of the axiom that if R and S are equivalent relations and x is a letter then the relation $\tau_x(R) = \tau_x(S)$ is true.

Set theorists often wish to form the class of all sets with some property: $\{x \mid \Phi(x)\}$; and a question whenever such a class is formed is whether it is a set. How does Bourbaki handle that?

Bourbaki has a notation for set formation; but it is introduced in an ambiguous way. I quote from page E II.4:

> *Trés fréquemment, dans la suite, on disposera d'un théorème*
> *de la forme $Coll_x R$ [defined on page E II.3 as $(\exists y)(\forall x)((x \in$*
> *$y) \Longleftrightarrow R)$, where y is a variable distinct from x and not*
> *occurring in R]. On introduira, alors, pour représenter le*
> *terme $\tau_y(\forall x)((x \in y) \Longleftrightarrow R)$ ... un symbole fonction-*
> *nel; dans ce qui suit, nous utilisons le symbole $\{x \mid R\}$; le*
> *terme correspondant ne contient pas x. C'est de ce terme*
> *qu'il s'agira quand on parlera de "l'ensemble des x tels que*
> *R". Par définition (I, p. 32) la relation $(\forall x)((x \in \{x \mid$*
> *$R\}) \Longleftrightarrow R)$ est **identique** à $Coll_x R$; par suite la relation*
> *R est **équivalente** à $x \in \{x \mid R\}$.*

I understand that to mean that the symbol $\{x \mid R\}$ is to be introduced in all cases as an abbreviation for the term $\tau_y(\forall x)((x \in y) \Longleftrightarrow R)$, whether or not the relation R is collectivising in x, for in that last sentence, "identique" means what it says, the relation $(\forall x)((x \in \{x \mid R\} \Longleftrightarrow R)$ and the relation $Coll_x(R)$ are actually the same relation; which is why I regret the remark cited above encouraging an *abus de langage*: if the proposed laxity for terms were extended to relations, chaos would result.

On the other hand, the *par suite* remark will hold only for collectivised R: take R to be $x = x$; the class of such x is not a set; as we saw above, every term will be some set, and therefore some x will not be in the term $\tau_y(\forall x)((x \in y) \Longleftrightarrow R)$.

B·13 Thus there is an acute difference between the normal use, in ZF and many other set theories, of the class-forming operator $\{\mid\}$ with the Church conversion schema $x \in \{x \mid R\} \iff R$ holding for all classes whether sets or not, and the Bourbaki treatment whereby, magically, conversion holds for a class if and only if that class is a set.

B·14 Now let $R\{x\}$ and $S\{x\}$ be two formulæ such that R and S are, provably in Bou54, not collectivising in x; for example $R\{x\}$ might be $x \notin x$ and $S\{x\}$ might be "x is a von Neumann ordinal" and we could use the Russell paradox in the one case and the Burali-Forti paradox in the other to establish the non-set-hood of these classes.

Write C_R for $(\forall x)((x \in y) \iff R)$ and C_S for $(\forall x)((x \in y) \iff S)$. Then Bou54 proves that for all y, C_R is false, and proves for all y that C_S is false; so it proves that $C_R \iff C_S$. Hence, by axiom S7, page E I.38, $\tau_y C_R = \tau_y C_S$: hence our notation is highly misleading in that all classes which are not sets have become "equal".

B·15 EXAMPLE As illustration, let T be a term which is certainly null, namely the universe. In Bourbaki's notation the following is a theorem:

B·16 THEOREM $\exists b\; b = \{x \mid x = x\}$.

Any ZF-iste reading that assertion would interpret it as false. The following is also a theorem of Bou54:

B·17 THEOREM $\exists a\; a = \{x \mid x \notin a\}$

and indeed the b of Theorem B·16 and the a of Theorem B·17 are equal.

Perverted interpretation of quantifiers

The step from an uninterpreted formal language to its possible interpretations is one of great epistemological importance. So, phenomenologically, the τ-operator perverts the natural order of mental acts: to interpret $\exists$ you look for witnesses and must first check whether a witness exists before you can pick an interpretation for $\tau_x(R)$; and then you **define** $(\exists x)R$ to mean that the witness witnesses that $R(\tau_x(R))$: a strange way to justify what Bourbaki, if pressed[30], would claim to be a meaningless text.

Discussion

B·18 Hilbert developed his operator in the belief that it would lead to consistency proofs for systems of arithmetic and of set theory; at the time he believed in the completeness of mathematics. In a complete system,

---

[30] See the remarks below in D·11 and the anti-ontologist jibe in issue 15 of *La Tribu* quoted in H·8.

such as the theory of a particular model, the operator works happily as a formal version of a Skolem function; but it starts to behave strangely in incomplete systems; and work in logic has shown that the phenomenon of incompleteness is not something marginal but pervades mathematics.[31]

B·19 A second count against Hilbert's operator is its blurring of subtleties concerning the various forms of the Axiom of Choice. If you then immerse yourself in a formalism based on it, how can you discuss models where AC is false? Indeed, the use of the τ-operator in this fashion seems to render impossible discussion of non-AC models, and as Gandy remarked in his review [G1] of *Théorie des Ensembles*, it makes it hard to tell when the Axiom of Choice is being used or not.

B·20 The strange way in which null terms behave makes the τ-calculus peculiarly unsuitable for use in set theory, where the idea of a proper class is important; and, of course, proper classes are important to the way category theorists think about functors; so that my objections to Bourbaki's logic apply both to a set-theory and to a category-theory based view of mathematics.

I am, I admit, being unfair to Hilbert in that I am writing eighty-nine years later, and benefitting from the enormous development of logic and set theory that has taken place in that time. But a modern account of foundational questions must allow for that development, and Bourbaki shut their eyes to it.

B·21 The idea of an expanding mathematical universe is essential to contemporary set theory. Consider the construction by Gödel of his inner model of AC and GCH. We have an iteration along the ordinals; at each successor stage we look at the class of those things definable over the previous stage; so the Bourbaki trick that a proper class "equals" some pre-assigned set is not the right mind-set for this construction. If you have immersed yourself in Bourbaki, Gödel, or set-theoretic recursion in general, will make no sense to you.

B·22 The "intuitive interpretation" where you choose once and for all witnesses for every true existential statement: implicit in that is the idea that you will choose one universe once and for all, which is precisely the notion that, according to Goldfarb, was current at the start of the nineteen-twenties. But if you are imbued with that idea, how are you to understand arguments, common in contemporary set theory ever since Cohen's announcement [Coh1], which involve passing from one universe to a larger one obtained by a forcing extension? If you have immersed yourself in

---

[31] The writings of Harvey Friedman give many examples; the paper [M5] describes four from analysis.

Bourbaki, Cohen will make no sense to you.

B·23 COMMENT So it might reasonably be said that Bourbaki's *Théorie des Ensembles* provides a foundation for many parts of mathematics, but not for set theory. That in turn provokes the thought that a mathematician who takes the Bourbachiste shilling *ipso facto* switches off that portion of his mathematical intelligence that can respond to the insights of set theory.

Bourbaki's remarks on progress in logic

Their historical remarks at the end of their set theory volume are well worth reading;□ it is only as their narrative comes up to modern times that one senses a pressure to adopt their view to the exclusion of all else. We quote various remarks that show that Bourbaki have a lively sense of logic as a developing subject.

They remark that philosophers do not have an up-to-date idea of mathematics and that logic is concerned with many things outside mathematics; but the neglect by philosophers of mathematics helps block progress in formal logic. They acknowledge that mathematicians with a good grasp of philosophy are equally rare.

On E IV.37, a footnote mentions the legendary event when at a lecture at Princeton in the presence of Gödel, the speaker said that there had been no progress in logic since Aristotle;* I sense a hint that the speaker might have been a philosopher. The speaker—and I would love to know who it was—was in good company: Bourbaki mention Kant's dictum that there was no need for new ideas in logic.

On page IV.42, they mention the harm done to Peano by Poincaré's often unjust jibes. The seven enthusiastic collaborators of Peano in the creation of his *Formulario Mathematico* of 1895 are listed.

B·24 They mention that Leibniz had the idea of Gödel numbering; that Frege had good ideas but bad notation; that Peano's notation was good; that Russell and Whitehead created a formalised language that "happily combines the precision of Frege with the convenience of Peano"; and that among subsequent modifications "the most interesting is certainly the introduction by Hilbert of" his symbol; they note that there was a shift to the dual; and that the symbol (apparently) bypasses the axiom of choice.

B·25 The paradoxes are discussed and the "Five letters" [Had] are mentioned, as is, on page 70, without naming him, Hermann Weyl's consequent resolve, quoted below in D·10, to avoid areas with paradoxes.

□ According to *La Tribu*, Weil was to consult Rosser when preparing the *Note Historique*.

* Might Rosser have been the source of this information?

B·26 They trace the history that culminated in the treatise of Hilbert and Bernays, with references to Poincaré, Ackermann, Herbrand and von Neumann, but, strangely, in their bibliography list Hilbert–Ackermann but not Hilbert–Bernays. Kleene's 1952 book [Kl] is listed; as (in the 1970 edition) are Cohen's PNAS announcements [Coh1].

Bourbaki's account of the incompleteness theorem

B·27 Their formulation of the incompleteness theorem is correct, if in their definition on page E I.21, a *theory* is to be a deductive system with finitely many explicit axioms and finitely many axiom schemes, (such as Zermelo's Aussonderungsprinzip), the substitution instances of which will be called the implicit axioms of the theory. Hence a theory, according to their definition, will be recursively axiomatisable. They put much emphasis on the role of substitution; they have to consider the result of substituting terms for variables in the explicit axioms, but need not do so for the implicit axioms, which by definition will form a class of wffs closed under the relevant substitutions.

B·28 Bourbaki define a formula to be *false* if its negation is provable. On page E IV 75, in footnote 1, they say that of any mathematical proposition it will eventually be known whether it is true, false or undecidable; so they are aware of the distinction between falsehood and unprovability. But their love of identifying truth with provability lands them in a tangle on Page E IV 74, in footnote 3: they say, incorrectly, that the proposition $S\{\gamma\}$ affirms its own falsehood, but it actually affirms its own unprovability.

B·29 COMMENT We shall see that this idea that systems are complete, with its corollary that "unprovable" is the same as "refutable", leads to repeated errors of logic in the works of members of the Bourbaki school.

C: ... *and by Godement, though with expressions of distrust.*

R OGER GODEMENT is listed by Cartier in an interview [Sen] in the
Mathematical Intelligencer as a member of the second generation
of Bourbaki, in the same group as Dixmier, Eilenberg, Koszul, Samuel,
Schwartz, and Serre. In 1963 his *Cours d'Algèbre* was published by Hermann of Paris. An English translation, copyrighted by Hermann in 1968,
was published by Kershaw of London in 1969, and runs to some 600 pages,
of which the first hundred are devoted to an account of logic and set theory.

I used the last 500 pages of this text for many years during the time
that I was teaching algebra to Cambridge undergraduates, and found it
excellent. When, later, I read the first part, on Set Theory, consisting of
§§0-5, I found, unhappily, that the account offered, although containing
many tart remarks to delight the reader, is flawed in various ways: one
finds errors of metamathematics, mis-statements of the results of Gödel
and Cohen, and an accumulation of negative messages about logic and set
theory.

In this chapter I shall present my findings by combing quickly through
his account and commenting as I go. I follow Godement in using the sign
§ to indicate chapter and N^o to indicate section within a chapter.

In the main he follows Bourbaki's *Théorie des Ensembles*. I have re-
arranged this material and simplified it in one or two places. Quotations,
from the English 1968 translation of his book, are given in *slanted type*;
the pagination follows that edition.

Godement has, pleasingly, in later French editions, moderated some of
the provocative statements on which I comment, and I shall in such places
draw attention to those revisions.

Godement's formal system

Godement states that the opening chapter is "an introduction to mathematical logic", and then adds, somewhat alarmingly, that "in it we have
tried to give a rough idea of the way mathematicians conceive of the objects
they work with." That opening chapter begins on page 20, where he writes:

> *In mathematics there are three fundamental processes: construc-*
> *tion of mathematical objects, the formation of relations between*
> *these objects, and the proof that certain of these relations are*
> *true, or, as usually said, are theorems.*
>
> *Examples of mathematical objects are numbers, functions,*
> *geometrical figures, and countless other things which mathemati-*
> *cians handle; strictly speaking these objects do not exist in Nature*

but are abstract models of physical objects, which may be complicated or simple, visible or not. Relations are assertions (true or not) which may be made about these objects, and which correspond to hypothetical properties of natural objects of which the mathematical objects are models. The true relations, as far as the mathematician is concerned, are those which may be logically deduced from a small number of axioms laid down once and for all. These axioms translate into mathematical language the most "self-evident" properties of the concrete objects under consideration; and the sequence of syllogisms by which one passes from the axioms (or, in practice, from theorems previously established) to a given theorem constitutes a proof of the latter.

C·1 COMMENT Notice the use of the phrase **once and for all**.

He continues:

Explanations of this sort, which may perhaps appear admirably clear to some beginners, have long since ceased to satisfy mathematicians: not only because mathematicians have small patience with vague phrases, but also and especially because mathematics itself has forced [mathematicians] to consider carefully the foundations of their science and to replace generalities by formulas whose meaning should be altogether free from ambiguity, and such that it should be possible to decide in a quasi-mechanical fashion whether they are true or not; and whether they make sense or not.

C·2 COMMENT Notice the residual faith in the completeness and decidability of mathematics implicit in that last sentence.

After some further discussion he begins his presentation of the formal system Bou54. Broadly he uses the syntax summarised in my Section B, but where Bourbaki speaks of terms, he speaks of mathematical objects; otherwise he follows Bourbaki's development, with, as in Bourbaki's first edition, a primitive sign, O, for the ordered pair of two objects. He immediately states that he will write (A, B) rather than $\mathsf{O}AB$.

He departs from Bourbaki by seeking to soften the austerity of the formalism, which leads him to make mistakes of logic, for he blurs the distinction between a uninterpreted formal language and its interpretations; and by a strange reluctance to state the axioms of set theory.

C·3 To introduce the axioms of logic, he writes, at the bottom of page 21:

Once the list of fundamental signs has been fixed, and the list of criteria of formation of mathematical objects and relations, it remains to state the axioms. Some will be purely logical, others of a strictly mathematical nature.

On page 25, he says that

> **true** *relations or* **theorems** ... *are those which can be obtained by repeated application of the two rules:*
> *(TR 1): Every relation obtained by applying an axiom is true*
> *(TR 2): if R and S are relations, if the relation $(R \implies S)$ is true and if the relation R is true, then the relation S is true.*

and on page 26, he says that

> *A relation is said to be* **false** *if its negation is true.*

and then immediately in Remark 1 that:

> *what characterizes true relations is that they can be proved.*

C·4 COMMENT This equation of truth and provability is the standard Bourbachiste position, found in the papers of Cartan and of Dieudonné of 1939-43 cited in *The Ignorance of Bourbaki* [M3], and found again in Dieudonné's last book [Di3, 1987, 1992]. But Godement then, in *Remark 2* cautions the reader with the symbol for a *tournant dangereux*:

> *There is a natural tendency to think that a relation which is "not true" must necessarily be "false". ... Unfortunately there is every reason to believe that in principle this is not so.*

We shall return to the rest of *Remark 2* later.

C·5 REMARK Thus words like "true" and "satisfies", as for example in the sentence

> *The mathematical object A is said to satisfy the relation R if the relation $(A|x)R$ is true,*

are being defined in terms of the provability relation $\vdash_{\text{Bou54}}$ that Godement is developing. Note, too, his comment in *Remark 4* of §0, on page 28:

> *[from (TL 2)] ... the relation R* **ou** (**non** *R) is true. It does not follow that at least one of the relations R,* **non** *R is true: this is precisely the question of whether there exist undecidable relations!*

C·6 Of (TL 7), which says that if $R(x)$ is true so is $(A|x)R$, Godement comments at the top of page 31 that

> *[its] purpose ... is precisely to justify this interpretation of letters as "undetermined objects";*

C·7 COMMENT We were told on page 21 that assemblies are built up from fundamental signs and letters, which suggests that a letter is a symbol, but at the top of page 30 we read

> *Let R be a relation, A a mathematical object, and x a letter (i.e. a "totally indeterminate" mathematical object),*

so by page 30 a letter is not a symbol: Godement is slipping away from treating his system as an uninterpreted calculus, and moving towards an informal Platonism.

C·8 Quantifiers are introduced on page 31 exactly as in Bourbaki, and need little comment beyond our remarks in Section B. He remarks on page 37 that

> we shall now be able, by using the Hilbert operation, to introduce them as simple abbreviations

— a phrase which in the light of the calculations of [M8] may strike the reader as a bit rich; and he ends the chapter with this pleasing remark:

> Like the God of the philosophers, the Hilbert operation is incomprehensible and invisible; but it governs everything, and its visible manifestations are everywhere.

Godement's set-theoretic axioms

Early in the next chapter, §1, he says

> ... if *a* and *b* are *mathematical objects* (or **sets**—the two terms are synonymous) ...

We shall see that that phrase causes trouble. A footnote on page 41 describes the less formal style he now wishes to adopt.

He introduces the axioms of equality in Theorem 1 on pages 41/42:

> ... intuitively this relation, when it is true, means that the concrete objects which *a* and *b* are thought of as representing are "identical". We do not enter into a philosophical discussion of the meaning of "identity" ...
>
> THEOREM 1: a) The relation $x = x$ is true for all x.
>
> b) The relations $x = y$ and $y = x$ are equivalent, for all x and all y.
>
> c) For all x, y, z, the relations $x = y$ and $y = z$ imply the relation $x = z$.
>
> d) Let u, v be objects such that $u = v$ and let $R\{x\}$ be a relation containing a letter x. Then the relations $R\{u\}$ and $R\{v\}$ ... are equivalent.

He comments that part d) is an axiom, while parts a), b) and c) can be deduced from

> a single much more complicated axiom, (which the beginner should not attempt to understand) namely that if R and S are equivalent relations and x is a letter then the relation $\tau_x(R) = \tau_x(S)$ is true.

C·9 COMMENT In Part d), formal system and informal interpretation are confused: otherwise "*such that $u = v$*" has no meaning. His definitions, taken *au pied de la lettre*, would imply that he means that if $\vdash_{\text{Bou54}} [u = v]$ then $\vdash_{\text{Bou54}} [R\{u\} \Longleftrightarrow R\{v\}]$, which is weaker than what I suspect he intended, namely that $\vdash_{\text{Bou54}} [u = v \Longrightarrow (R\{u\} \Longleftrightarrow R\{v\})]$—a point with consequences for the notion of *cardinal number*.

C·10 COMMENT It may be for pedagogical reasons that he introduces the axioms one by one, but it is regrettable that there is no signal to the reader when the presentation is complete.

Nothing is said about the origins of the axioms: they are presented as oracular pronouncements.

So far as I can tell, searching the first hundred pages and assuming that no further axioms will be introduced once he starts on the algebra, his set-theoretic axioms are these:

First, the axiom of **extensionality**, which appears on page 42:

... *In fact there is only one axiom governing the use of $\in$, namely:*
THEOREM 2: *Let A and B be two sets. Then we have $A = B$ if and only if the relations $x \in A$ and $x \in B$ are equivalent.*

I take the preamble to Theorem 2 to mean that he considers it to be an axiom: thus we have had one set-theoretic axiom so far, that of extensionality, but stated as a theorem.

C·11 COMMENT Note the further slipping between languages: "*the relations are equivalent*" means that a certain formula is provable in the system Bou54 we are building up. That is a clear assertion, and is said to hold iff $A = B$; but $A = B$ is an (uninterpreted) relation. Does he mean "$A = B$ is true iff the relations are equivalent", or does he mean "it is provable in Bou54 that $(A = B \Longleftrightarrow (x \in A \Longleftrightarrow x \in B))$"? That he is aware of the difference is evident from part (2) of Remark 7 on page 34.

C·12 COMMENT Theorem 2 as stated is false. Let x and y be distinct letters. Looking ahead to pages 46–7, where Godement adds the axiom of pairing, let A be $\{x\}$ and let B be $\{x, y\}$. I believe that A and B are mathematical objects; indeed a definition using τ can be given, though Godement does not do so: I assume that formally he would put $\{x\}$ $=_{\text{df}}$ $\tau_y((\forall z)((z \in y) \Longleftrightarrow z = x))$ and $\{x, y\}$ $=_{\text{df}}$ $\tau_z((\forall w)((w \in z) \Longleftrightarrow w = x \text{ ou } w = y))$. On page 41 he says that the terms *mathematical object* and *set* are synonymous. Very well, A and B are sets, and the following is provable in Bou54: $x \in A \Longleftrightarrow x \in B$, so that the relations $x \in A$ and $x \in B$ are equivalent. If Theorem 2 were provable, we could infer that $A = B$ and thus that $x = y$; so we would have proved that any two sets are equal!

Of course, what is lacking is a requirement that the letter x has no occurrence in A or in B. That slip is strange in that in the statement of Theorem 4 on page 44 (a version of the scheme of separation) he is careful to say that the relation R is to contain a variable x: though there the theorem would still be true if it did not. On the other hand the further statement of Theorem 4, that *for every set X there exists a unique subset A of X,* ... shows further confusion of language; *"for every set X"* might mean "for every mathematical object", that is "for every term ", but *"there exists a unique subset"* certainly is using "there exists" mathematically; A is a variable here not a term, and will be the subject of various assertions.

C·13 Of his version of the scheme of **separation** given in Theorem 4 of §1, on page 44, Godement says in Remark 4 on that page that

> *Mathematically, Theorem 4 cannot be proved without using axioms which are far less self-evident, and the beginner is therefore advised to assume Theorem 4 as an axiom*

so, so far, we have had extensionality and separation.

C·14 Then Godement, wishing to comment on the necessity of giving a scheme of separation rather than of comprehension, writes on page 44:

> *Remark 5: In spite of the dictates of common sense, it is not true that for every relation $R\{x\}$ there exists a set (in the precise sense of §0) whose elements are all the objects x for which $R\{x\}$ is true.*

C·15 COMMENT When I turn to §0, I find that the word "set" occurs only thrice, namely on page 20 in N°1, in the phrases *the development of the "theory of sets", Set theory was created by Cantor,* and *set theory had given rise to genuine internal contradictions;* I can find it nowhere else in §0, although there is a lot of talk about **objects**. The word **sets** appears, in bold face, early in §1, suggesting that that is its definition, namely that a set is the same as a mathematical object. So, as I can find no precise sense of "set" in §0, whereas "mathematical object" is given a very precise sense in that section, namely that it is a term in a certain carefully specified formal language, and as I am told that a set is the same as a mathematical object, so be it: I shall take 'set' to mean 'term in Godement's formal language'.

Remark 5 continues:

> *Suppose that there exists a set A such that the relations $x \in A$ and $x \notin x$ are equivalent.*

He hopes to get a contradiction, but there is nothing wrong with that, as it stands: let A be

$$\tau_y((\forall z)((z \in y) \Longleftrightarrow (z = x \,\&\, z \notin z))).$$

Then A is a mathematical object, therefore a set, and the relation $x \in A \iff x \notin x$ is true.

C·16 REMARK Again, what is missing is the requirement that the mathematical object A have no occurrence of the letter x; then his remark that the supposition of the existence of such an A leads to contradiction would become correct.

C·17 We shall refer below to his Remark 6, on page 45, in which he says that apparently obvious assertions cease to be so simple when it is a question of effectively *proving* them.

The empty set is discussed in $N^{o}4$: if there is a set then the empty set exists, by the scheme of separation: no axiom so far asserts the unconditional existence of any set; but the τ-formalism guarantees that the existence of something is provable.

Godement goes on in $N^{o}5$ to discuss sets of one and two elements, and then says *In the same way we can define sets of three, four ... elements. The sets so obtained are called finite sets, and all other sets are called infinite sets. These two notions will be considered afresh in §5.*

C·18 COMMENT Note that that is not a formal definition of "finite", the use of dots constituting an appeal to the reader's intuitive notion of finiteness.

C·19 In §1, $N^{o}5$, Remark 8, on page 47, he states that the existence of the **pair set** of two objects is an axiom, and goes on to say that the existence of **infinite sets** is also an axiom, and acknowledges that we have yet to define the natural numbers, which will be done in §5; so far we have had axioms of extensionality, separation, pairing and infinity, but we await a definition of *finite*.

In $N^{o}6$ the **set of subsets** of a given set is introduced thus:

Let X be a set. Then there exists — this again is one of the axioms of mathematics — one and only one set, denoted by

$$\mathcal{P}(X)$$

with the following property: the elements of $\mathcal{P}(X)$ are the subsets of X,

so hitherto we have had extensionality, separation, pairing, infinity, and power set; we await a definition of *finite*.

C·20 In §2 he discusses ordered pairs and Cartesian products, no new axioms being introduced till §3. In Remark 1 of §2 on page 50, he writes:

The [Kuratowski method of defining] ordered pair is totally devoid of interest. ... The one and only question of mathematical importance is to know the conditions under which two ordered pairs are equal.

C·21 COMMENT To an algebraist, that might be true. But to a set-theorist interested in doing abstract recursion theory, it is very natural to ask whether a given set is closed under pairing. For that reason, an economical definition of ordered pair is desirable, such as is furnished by Kuratowski's definition: otherwise one might find that the class of hereditarily finite sets is not closed under pairing, or even that no countable transitive set is. Bourbaki in their later editions have indeed adopted the Kuratowski ordered pair.

In §2 N°2 he declares that
using the methods of §0, cartesian products can be proved to exist.

C·22 COMMENT I deny that that can be proved from the axioms he has stated so far, given that he has refused to define ordered pair — hence we do not know where the values of the unpairing functions (projections) lie — and he has not stated a scheme of replacement.

Write $\langle x, y \rangle_K$ for the Kuratowski ordered pair $\{\{x\}, \{x, y\}\}$, and $u \times_K v$ for $\{\langle x, y \rangle_K \mid x \in u \ \& \ y \in v\}$, the correspondingly defined Cartesian product.

Let "F defines a possible pairing function", where F is a three-place Bourbaki relation, abbreviate the conjunction of these statements:

(C·22·0) $\forall x \forall y \exists$ exactly one z with $F(x, y, z)$

(C·22·1) $\forall z \forall u \forall v \forall x \forall y \big(F(u, v, z) \ \& \ F(x, y, z) \implies [u = x \ \& \ v = y] \big).$

Let "$X \times_F Y \in V$" denote the formula $\exists W \forall w (w \in W \iff \exists x \exists y (x \in X \ \& \ y \in Y \ \& \ F(x, y, w)))$.

Consider the following argument, which I present in ZF-style set theory.

C·23 LEMMA *Suppose $x \mapsto G(x)$ is a one-place function with domain V. Define $F(x, y, z) \iff_{\mathrm{df}} z = \langle G(x), \langle x, y \rangle_K \rangle_K$. Then F defines a possible pairing function.*

Proof : Write $(x, y)_F$ for $\langle G(x), \langle x, y \rangle_K \rangle_K$. We have only to check that the crucial property $(x, y)_F = (z, w)_F \implies x = z \ \& \ y = w$ is provable. But that is immediate from the properties of the Kuratowski ordered pair.
$\dashv$ (C·23)

C·24 LEMMA *Now let A be a set, and let $B = \{\varnothing\}$; let G and F be as above; then if $A \times_F B \in V$, then the image $G``A$ of the set of points in A under G is a set.*

Proof : If $A \times_F B \in V$,

$$A \times_F B = \{(a, b)_F \mid a \in A, b \in B\} = \{\langle G(a), \langle a, b \rangle_K \rangle_K \mid a \in A, b \in B\}.$$

$G``A = \{G(a) \mid a \in A\} \subseteq \bigcup\bigcup(A \times_F B)$, so the lemma will follow from two applications of the axiom of union, which will be become available to us on a second reading of §3 N°2, and an application of the scheme of separation, which we have been advised to take as an axiom.   ⊣ (C·24)

Now let **GT** be the system comprising the axioms of extensionality, pairing and power set, together with the scheme of separation: since $\vdash_{GT}$ $u \times_K v \subseteq \mathcal{P}(\mathcal{P}(\bigcup\{u, v\}))$, the set-hood of $u \times_K v$ when u and v are sets can be proved in **GT** plus the axiom that $u \cup v \in V$, a weak form of the axiom of union.

C·25 METATHEOREM *The following systems are equivalent over* **GT***:*

(C·25·0)   *Bourbaki's scheme S8 of selection and union;*

(C·25·1)   *Godement's scheme of union, discussed below;*

(C·25·2)   *the axiom of union plus the scheme of replacement;*

(C·25·3)   *the axiom of union plus the scheme that for each formula F with three free variables, the sentence expressing "if F defines a possible pairing function then for each A and B, $A \times_F B$ is a set" is an axiom.*

Since **GT** is a subsystem of **Z**, in which many instances of replacement fail, Godement's claim to prove the existence of cartesian products, no matter what definition of ordered pair is used, must also fail.

C·26 In Remark 4 on page 56, the existence of the set of natural numbers is assumed (and referred to the existence of infinite sets, stated to be an axiom, but not formulated: so far there has been no proper definition of *finite*).

C·27 The **Axiom of Choice** sneaks in *via* the τ-operator, about which Godement has said, on page 37, §0 N°9:

> It is also used nowadays in place of the Axiom of Choice. (§2 Remark 7).

Turning to §2 N°8, on page 63, we find in Remark 8, not 7, a demonstration of the use of τ to get **AC**:

> ... for want of anything better we can define a function h by $h(y) = \tau_x(f(x) = y)$ ...

C·28 In §3 he turns to unions and intersections. In Remark 1 on page 70 it is mentioned that we need an axiom of union to form $X \cup Y$, with forward reference to N°2. Of N°2 it is said that it may be omitted at a first reading. It is stated that the existence of the **union** of an arbitrary family is an axiom of mathematics. But **ZF**-istes must beware! for Godement's axiom of union is really a scheme which is much stronger than the simple axiom of union, and therefore I shall speak of it as Godement's scheme of union.

With that point in mind, we have now had the axioms of extensionality, pairing, infinity, and power set, and the schemes of separation and union; we continue to await a definition of *finite*.

C·29 That his system is equivalent to Bourbaki's follows from an examination of Bourbaki's *schéma de sélection et réunion* found as S8 of the *Théorie des Ensembles*, on page E II.4:

> Soient R une relation, x et y des lettres distinctes, X et Y des lettres distinctes de x et y et ne figurant pas dans R. La relation

$$(\forall y)(\exists X)(\forall x)(R \implies (x \in X)) \implies (\forall Y)\mathrm{Coll}_x((\exists y)((y \in Y) \text{ et } R))$$

> est un axiome.

Bourbaki is aware of the power of S8, drawing attention to the difference between the union of a family of subsets of a given set and the union of a family of sets where no containing set is known. Thus the ZF-ists' axiom of union together with schemes of separation and replacement is equivalent to Godement's scheme of union and to Bourbaki's scheme of selection and union.

C·30 §4, on equivalence relations, calls for little comment. Example 4 on page 78 speaks of the set of rational integers, though we still await a definition of ℕ.

C·31 In §5, on *Finite sets and integers* the concept of finiteness will be at last defined. Kronecker's witticism is here attributed to Dedekind. Godement remarks that *the integers with which we are concerned here are mathematical objects, not concrete integers*, which underlines the need for a formal definition of *finite*.

In N°1, he introduces the notion of equipotence; his Theorem 1 is a version of AC: any two sets are comparable. (Bernstein's proof is given in Exercise 5).

C·32 In §5 N°2, the cardinal of X is elegantly defined as $\tau_Y(Eq(X, Y))$, emphasizing the reliance on the identity of τ-selected witnesses to equivalent propositions: my readers should bear C·9 in mind.

On page 90, discussing equipotence, he emphasizes that the "ordinary" numbers are metaphysical ideas derived from concrete experience, whereas "Mathematical" numbers are objects defined by following the procedures of §0.

His treatment of cardinals follows Bourbaki. Thus 0 is defined on page 90 to be the cardinal of the empty set, therefore some object equipollent to the empty set; therefore (as remarked by Bourbaki but not by Godement) the empty set itself.

1 is the cardinal of the singleton of the empty set, so is some object with exactly one element and therefore not equal to 0. The paper [M8] shows how this definition gets out of hand.

2 is the cardinal of the von Neumann ordinal 2, so is some object with exactly two elements. The calculations of [M8] would presumably yield even more monstrously long assemblies for this and other finite cardinals. There is a forward reference to §5 N°4. Finally he assumes without proof:

Theorem 2: any set of cardinal numbers has a sup and an inf.

C·33 COMMENT That too involves an appeal to replacement. Without it, curious things happen. Suppose we define $Card(n)$ to be the set $\{\aleph_k \mid k < n\}$, where we take $\aleph_k$ to be an initial ordinal: a reasonable definition, by Bourbachiste standards, as it is indeed a set of cardinality n. But then in a set theory without some version of the axiom of replacement, (for example, in Zermelo set theory) the class of finite cardinals as we have just defined them need not be a set.

C·34 COMMENT There will be other unheralded uses of replacement: for example, the construction on page 97 of the set $\bigcup_{n\in\mathbf{N}} X_n$, and in the footnote on the same page, the proof that the class of cardinals less than a given cardinal is a set.

His exercise 3 page 108 asks the reader to prove that the countable union of countable sets is countable: another covert use of AC!

C·35 In §5 N°3, he defines the sum of a family of cardinals. That enables him, in N°4 on page 95, to follow Dedekind's approach and define a cardinal x to be *finite* if $x \neq x + 1$; a *natural number* is then defined to be a finite cardinal; and a set is finite if its cardinal is finite.

C·36 In N°5 Godement derives the existence of the set of natural numbers from the existence of infinite sets, which he has previously stated to be one of the axioms of mathematics. Thus his axiom of infinity would state that there is a cardinal x which equals $x + 1$.

C·37 REMARK His definition of *finite* relies on a mild form of the Axiom of Choice to be correct; a set is Dedekind-infinite iff there is an injection of $\mathbb{N}$ into it; and if ZF is consistent, then there are models of the ZF axioms without AC in which there are Dedekind-finite sets not equipotent to any finite ordinal.

C·38 That appears to be all the axioms given by Godement. I find no mention of an axiom of foundation; but, as we have seen, the scheme of replacement is embedded in his scheme of union.

C·39 COMMENT Mathematicians, wrote Godement, have small patience with vague phrases; why then should they tolerate the purposeless imprecision of the Bourbaki–Godement treatment of finite cardinals and AC?

Misunderstandings of work of logicians

When Godement comes to the work of Gödel and other logicians, he makes more serious errors. For example, on page 26, §0, N°4, Remark 2, after defining the notion of an *undecidable* relation, he wrote:

> At the present time no example is known of a relation which can be proved to be undecidable (so that the reader is unlikely to meet one in practice ...) But on the other hand the logicians (especially K. Gödel) showed thirty years ago that there is no hope of eventually finding a "reasonable proof" of the fact that every relation is either true or false; and their arguments make it extremely probable that undecidable relations exist. Roughly speaking the usual axioms of mathematics are not sufficiently restrictive to prevent the manifestation of logical ambiguities.

C·40 COMMENT Notice the use of the word **fact**.

C·41 COMMENT The French original ... *du fait que toute relation est soit vraie soit fausse* ... of those words was written in 1963, thirty three years after the incompleteness theorems were announced. What is one to make of these statements? If he believes that theorems are deduced from a small number of axioms laid down once and for all, and if that means that the set of axioms is recursive, then if his chosen system is consistent, undecidable relations are certainly known; and given his definition of "true" as provable and "false" as refutable, it is simply not the case that every relation is either true or false. Why is he so reluctant to allow Gödel's discoveries to be established rather than be merely "extremely probable"?

The last sentence quoted above is perfectly correct, more so than perhaps Godement realised. Incompleteness pervades mathematics: the phenomenon may be found in almost any branch of mathematics and is not something confined to artificial and contrived assertions on the very margin of our science.

C·42 In §0 N°4, on page 26, he writes

> Remark 3: ... contradictory relations are both true and false. The efforts of the logicians to establish a priori that no such relations exist have not so far met with success.

COMMENT That last statement wholly misrepresents the import of the Incompleteness Theorems.

C·43 We have seen that he follows Bourbaki in using τ to get the Axiom of Choice. His comment,

> The possibility of constructing a [choice set] (which is obvious if one uses the Hilbert operation) is known as the Axiom of Choice.

> *Until recent times it was regarded with suspicion by some mathe-*
> *maticians, but the work of Kurt Gödel (1940) has established that*
> *the Axiom of Choice is not in contradiction to the other axioms*
> *(which of course does not in any way prove that the latter are*
> *non-contradictory)*

is well-phrased, though it might leave the reader wondering what the problem was that Gödel solved.

C·44 The first French edition of Godement's *Cours d'Algèbre* was written,
presumably, just before the announcement of Cohen's discoveries. In Remark 9 of §5 N°7, on page 98 of the 1969 English translation, he mentions
Cohen's "magnificent result" that CH is undecidable, and acknowledges
that his earlier assertion that the reader would not encounter an undecidable relation should now be amended; but he makes no mention of Gödel's
consistency proof for CH.

In his later French editions, such as the impressions of 1973 and 1980,
he does mention Gödel's relative consistency proof for the continuum hypothesis, and also rewrites the other three Remarks that I have just quoted,
so that his readers are now told *something* about the existence of undecidable statements in normal mathematics.

Unease in the presence of logic

Besides the above mis-statements by Godement concerning the work
of logicians, we find repeatedly an undertow of unhappiness about logic,
which, sadly, has not been corrected in his later French editions.

He writes on page 22, in §0 N°2,

> *It has been calculated that if one were to write down in formal-*
> *ized language a mathematical object so (apparently) simple as*
> *the number 1, the result would be an assembly of several tens of*
> *thousands of signs.*

C·45 COMMENT This remark goes back to Bourbaki, and it is shown in
[M8] that the estimate given is too small by a factor of perhaps a hundred
million. But even if their estimate were correct, **what is the point of all
those symbols?** Why not follow Zermelo and von Neumann and define 0
as ∅ and 1 as {0}?

Poincaré mocked Couturat for taking perhaps twenty symbols to define
the number 1, in an attempt to reduce that arithmetical concept to one of
logic; now Bourbaki is taking 4 European billions (= American trillions) of
symbols to do the same thing: one million thousand-page books of densely
packed symbols. Suppose an error occurred somewhere in those pages:
would anyone notice? Would it matter? That is not where the mathematics
resides.

C·46 In short, the chosen formalism is ridiculous, and Godement knows that it is ridiculous, for he makes the excellent remark:

> *A mathematician who attempted to manipulate such assemblies of signs might be compared to a mountaineer who, in order to choose his footholds, first examined the rock face with an electron microscope.*

Did it ever occur to him to wonder whether other formalisms might be possible? I fear that where logic is concerned, Godement's state of mind is that imputed[32] by Strachan-Davidson to the schoolboy who "believes in his heart that no nonsense is too enormous to be a possible translation of a classical author."

Let us list the other symptoms:

On page 25, there is a footnote:

> *it is very difficult, in practice, to use the sign* $\Longrightarrow$ *correctly.*[v]

On page 31, another footnote:

> *it is very difficult to use the signs* $\exists$ *and* $\forall$ *correctly in practice, and it is therefore preferable to write "there exists" and "for all", as has always been done.*[vi]

What will he do, I asked myself, with the set-forming operator? The answer astonished me: he does not use it. I have been right through the book searching, and I cannot find it at all. He introduces signs for singletons and unordered pairs; but every time he wants to introduce a set, for example a coset in a group, he writes out in words "let F be the set of ....".

In Remark 5 of §1 N°3, on page 44, Godement says

> *these examples show that the use of the word "set" in mathematics is subject to limitations which are not indicated by intuition.*[vii]

In Remark 6 he says

> *... apparently obvious assertions cease to be so simple when it is a question of effectively proving them. The Greeks were already*

---

[32] according to Sir Donald Francis Tovey [T, page 10].

[v] *Il est du reste fort difficile, dans la pratique, d'utiliser *correctement* le signe* $\Longrightarrow$.

[vi] *Il est fort difficile d'utiliser *correctement* les signes* $\exists$ *and* $\forall$ *dans la pratique courante; il est donc préférable de se borner à écrire "il existe" et "pour tout" comme on l'a toujours fait.*

[vii] *Ces exemples montrent que l'usage du mot "ensemble" est soumis en Mathématiques à des limitations que l'intuition n'enseigne pas.*

aware of this.[viii]

and on page 98, in commenting on Example 1 of §5 N°5, he says:

> *it is precisely one of Cantor's greatest achievements that he dis-*
> *qualified the use of "common sense" in mathematics.*[ix]

C·47 COMMENT The cumulative effect of all these comments is this: Godement tells the reader that a simple concept such as the number 1 can take thousands of signs to write out formally, that it is very difficult to use connectives correctly, and that it is very difficult to use quantifiers correctly. Coupled to this comprehensive group of negative messages about logic are some equally discouraging statements about set theory: that the concept of 'set' is counter-intuitive, that apparently obvious assertions are hard to prove, that common sense has been disqualified from set theory; further, he avoids the usual notation for forming sets and he evinces a remarkable reluctance even to state the axioms of set theory.

Can I be blamed for suspecting that Godement distrusts formalised reasoning? I know he says, in §0 N°1, on page 22, that

formalized mathematics exists only in the imagination of mathematicians[x]

but I feel he would rather even that did not happen. He brings to mind a remark of Padoa:

> *Logic is not in a good state: philosophers speak of it with-*
> *out using it, and mathematicians use it without speaking*
> *of it, and even without desiring to hear it spoken of.*

In sum, his message is that logic and set theory are a morass of confusion: but what has happened is that Bourbaki, whom he follows, have chosen a weird formalisation, they have noticed that in their chosen system proof is very awkward, and they have concluded that the whole thing is the fault of the logicians.

Nowhere, in Bourbaki or in Godement, is there any suggestion that other formalisations are possible. Godement says "mathematicians are impatient of vague statements", he explains that formality is a good thing, and then like a sharper forcing a card, offers you a choice of exactly one formalisation, and, at that, one that is cumbersome and destructive of intuition.

[viii] *Ceci montre que des assertions en apparence évidentes cessent d'être simples lorsqu'on veut effectivement les démontrer, c'est ce que les Grecs avaient déjà remarqué.*

[ix] *C'est précisément l'une des plus grandes réussites de Cantor que d'avoir pu disqualifier d'emploi du "bon sens" en Mathématiques.*

[x] *Elles n'existent bien entendu que dans l'imagination des mathématiciens.*

C·48 If their chosen system, Bou54, is what the Bourbachistes think logic and set theory are like, it is no wonder that they and their disciples are against those subjects and shy away from them. But on reading through Godement one last time, I was left with the impression that he is not so much a disciple of Bourbaki as a victim: loyalty to the group has obliged him to follow the party presentation of logic and set theory, and his intelligence has rebelled against it. I would love to teach him.

D: *It is this distrust, intensified to a phobia by the*
     vehemence of Dieudonné's writings, ...

JEAN DIEUDONNÉ is acknowledged by many witnesses to have had a central, even a dominant, rôle in the successful functioning of the Bourbaki group: Armand Borel in his essay [Bor] mentions shouting matches, generally led by Dieudonné with his stentorian voice, and writes

> "There were two reasons for the productivity of the group: the unflinching commitment of the members, and the superhuman efficiency of Dieudonné."

Pierre Cartier in his interview [Sen] with the *Mathematical Intelligencer* describes Dieudonné as "the scribe of Bourbaki", and also makes, among numerous thoughtful points, these significant assertions:

> **Bourbaki never seriously considered logic.**
> **Dieudonné himself was very vocal against logic.**

D·0 That last disclosure is endorsed by a passage in Quine's autobiography [Q, page 433]:

> "[In 1978] a Logic Colloquium was afoot in the École Normale Supérieure. [...] Dieudonné was there, a harsh reminder of the smug and uninformed disdain of mathematical logic that once prevailed in the rank and vile, one is tempted to say, of the mathematical fraternity. His ever hostile interventions were directed at no detail of the discussion, which he scorned, but against the enterprise as such. At length one of the Frenchmen asked why he had come. He replied 'J'étais invité.' " [33]

There is admittedly a tradition of hostility to logic in France: one can look back to the suppression of Port Royal in the 17th century, and yet earlier to Abélard who lamented that his logic had made him odious to the world; in the twentieth century, Poincaré[34] quipped that "la logique n'est plus stérile, elle engendre des paradoxes"; Alexandre Koyré in "Epiménide le menteur"[35] wrote, less charitably: "la logique symbolique forme une discipline hybride, aussi ennuyeuse que stérile"; and papers of Tarski on the Axiom of Choice submitted to the *Comptes Rendus* in the late 1930s

---

[33] An eye-witness has suggested that Quine might have been over-reacting to Dieudonné's characteristic behaviour.

[34] See Chapter II of [M-K] for some consequences for France of Poincaré's perceived position on logic.

[35] Reviewed by Max Black in [Bl2].

were rejected by Lebesgue as absurd and then by Hadamard on the grounds that AC is a trivially obvious fact.

But it might be said that Bourbaki have continued that tradition, perhaps taking their cue from Poincaré without remarking that he was against logicism rather than against logic. We have seen the undertow in Godement's treatise; and if we look for further evidence of the antipathy of the Bourbachistes to logic, we find that the finger points at Dieudonné. So whether or not the Bourbaki bias was due solely to one extremely energetic man, it seems desirable to examine Dieudonné's position.

But that is easier said than done, for just as Leibniz has been spotted (notably by Couturat, following Vacca) saying one thing to the Queen of Sweden and saying another to his private diary, Dieudonné seems to have one opinion when he is thumping the tub on behalf of Bourbaki and another when he is musing as a private individual. Though his energy is evident, the coherence of his position is not.

Dieudonné's earlier writings on foundational questions have been touched on in [M3]; here we look at his essay entitled *La Philosophie des mathématiques de Bourbaki* which is to be found on pages 27-39 of Tome I [Di2] of an anthology of his papers in two volumes published by Hermann of Paris in 1981 with the title *Choix d'Œuvres de Jean Dieudonné de l'Institut*.

The essay contains many good debating points and delightful jibes at various figures such as Russell and (to my surprise) Poincaré, but the title is strange, since it suggests that the essay will present or discuss the philosophy of mathematics of Bourbaki, yet the author is at pains to explain that Bourbaki have only ever made two statements about the philosophy of mathematics, and that the opinions expressed are his own.

I could find no reference, in that anthology, to any previous appearance of that essay, which on internal evidence seems to be the text of a talk given at, and towards the end of, a gathering of philosophers held not before 1977, since he cites his *Panorama des mathématiques pures*, which was published in that year.

D·1 The first message of the paper is to warn his audience that philosophers of mathematics are unaware of the scope of modern mathematics, and whilst they believe they are discussing today's mathematics, in fact they are discussing that of the day before yesterday.

D·2 COMMENT That chimes with my experience of joint seminars on this theme: the philosophers know no mathematics and the mathematicians no philosophy, so that the interaction is weak. My readers should be similarly warned that Dieudonné is not discussing logic and set theory as understood in the first decade of the twenty-first century, but the logic and set theory of the first quarter of the twentieth.

D·3 So now Dieudonné will tell the philosophers what mathematicians are up to, and his second message is that 95% of mathematicians agree with Bourbaki and the rest are crazy, and arrogant with it.

He has two ways of saying "almost all" mathematicians: one is to say that 95% of mathematicians do this or think that; and the other is to say that the quasi-totality of mathematicians do that and say this. I am not sure whether the quasi-totality means at least 99.9%.

D·4 He considers that whereas at the beginning of the twentieth century foundational questions commanded the attention of many of the greatest mathematicians of the day, from 1925 onwards that has ceased to be the case, and that logic and set theory have become marginal disciplines for the quasi-totality of mathematicians:

> *"Au début du vingtième siècle .. les plus grands mathématiciens se passionaient pour les questions des "fondements" des mathématiques; aujourd'hui le divorce est presque total entre "logiciens" et "mathématiciens". .. Il ne faut pas cesser de redire que, pour la quasi totalité des mathématiciens d'aujourd'hui, la logique et la théorie des ensembles sont devenues des disciplines marginales: elles se seraient définitivement arrêtées après 1925 qu'ils ne s'en apercevraient même pas.*
>
> *... je ne parle pas d'opinions mais de faits. Les travaux de Gödel, Cohen, Tarski, J. Robinson et Matijasevich n'exercent aucune influence."*

Dieudonné makes both his position, and his ignorance of developments in logic, very clear. His choice of the date 1925 is puzzling: does he allude to the completion of Ackermann's thesis?

D·5 He pauses to swipe at the study of large cardinals.

> *"Les spéculations sur les "grands" cardinaux ou ordinaux laissent froids 95% entre eux, car ils n'en rencontrent jamais."*

Let me pause too, to remark that the paper [M5] gives examples to show that there are straightforward problems of ordinary mathematics in which large cardinal properties prove to be embedded, even if ordinary mathematicians are not aware of the fact.

D·6 He warns that many philosophers unconsciously identify two parts of mathematics, logic-and-set-theory on the one hand, and the rest on the other,* which, he believes, are in fact strongly separated in the practice of mathematics: paragraph I 3 of the essay reads, in part:

---

* I take that as an admission that logic-and-set-theory are indeed part of mathematics.

le divorce est presque total entre les mathématiciens s'occupant
de Logique ou de Théorie des ensembles (que j'appellerai pour
abréger "logiciens") et les autres (que j'appellerai simplement "mathématiciens", pour ne pas toujours dire "mathématiciens ne s'occupant pas de logique ni de théorie des ensembles.")

D·7 Dieudonné then says he will give special meanings to the words "logicians" and "mathematicians"; I shall indicate those by capital letters;
thus he defines Logicians to be mathematicians concerned with logic or set
theory; and Mathematicians to be mathematicians-concerned-with-neither-
logic-nor-set-theory.

Presumably from that point on in his essay, he intends those words to
be used in that exclusive sense, and supposes that every mathematician is
either a Mathematician or a Logician, but not both.

But how exclusive is it? A mathematician who works in several areas,
one of which is logic, need not have logic on the brain the whole time.
Littlewood, for example, once wrote a paper on cardinal arithmetic, but no
one has ever accused him of being a logician. Everyone would call Shelah
a logician; but when he solved Kuroš' problem by proving that there is
an uncountable group with no uncountable subgroup, he harnessed small
cancellation theory to ideas from combinatorial set theory. And when he
gave primitive recursive bounds for van der Waerden's theorem, was he
doing number theory or was he doing recursion theory?

How, for example, would Dieudonné have categorized Louveau?

D·8 **How, indeed, would he categorize himself?** On pages 354–356
in the same volume [Di2] of Selected Papers we find a short note called
"Bounded sets in (F)-spaces", first published in the Proceedings of the
American Mathematical Society **6** (1955) 729–731.

He asks two questions about locally convex metric spaces, and **using
the Continuum Hypothesis** – ho, ho, ho – gives a counterexample to
each.**

He writes "It would be interesting to give negative answers without the
continuum hypothesis". Without looking at the specific questions asked,
we might remark his implied belief that a positive answer without the continuum hypothesis would not be interesting. Indeed it might place the
question beyond mathematics, for it would then not have been answered
within the scope of ZF.

Suppose his question had been equivalent to CH: would that be of

---

** He uses a lemma, assuming the Continuum Hypothesis, the conclusion of
which is easily equivalent to the assertion that $\mathfrak{d}$, the dominating number, equals
$\aleph_1$.

interest? Or is it that CH to him is only a prop for a pre-proof which, he hopes, will eventually discard that prop?

In his essay he declares that for the quasi-totality of today's Mathematicians logic and set theory have become marginal disciplines. — How did he view his own use of the continuum hypothesis? He thought his paper worth including in a selection made 26 years after the paper was originally published.

D·9 We have heard Dieudonné say that Gödel, Cohen, Tarski, Julia Robinson and Matijasevic, profound though their work is, have exerted no influence (positive or negative) on the solution of the immense majority of problems which interest Mathematicians.

Is that true? What one finds on surveying the mathematical scene is a slippery characteristic: a problem, thought to be Mathematical, which proves to have a solution using ideas from Logic, is liable to be declared by those formerly interested in it to have revealed itself to be unimportant. One might cite here the response of the late J. Frank Adams to the information that Shelah had solved Whitehead's problem concerning free Abelian groups: "Whitehead would only have been interested in the countable case".♣

There are many problems stated by Mathematicians to be of interest and which are later proved, usually by Logicians, to have a Logical component. The continuum hypothesis was included by Hilbert — a Mathematician or a Logician in Dieudonné's eyes? — in his famous list of 1900, and so was the search for an algorithm for the solution of Diophantine equations. What is this strange interest that is extinguished should the problem prove not to have an answer of the kind simple Mathematicians seem to expect?

D·10 Dieudonné claims to be speaking of facts; indeed he might almost be claiming, in Sir Herbert Butterfield's immortal phrase, that the 'facts' are being allowed to 'speak for themselves'. But the facts are not that clear. On pages 4 and 5 of their 1958 book on the *Foundations of Set Theory*, (and on page 4 of the revised edition [Fr-bH-L]), Fraenkel and Bar-Hillel write that "Nevertheless, even today the psychological effect of the antinomies on many mathematicians should not be underestimated. In 1946, almost half a century after the despairing gestures of Dedekind and Frege, one of the outstanding scholars of our times made the following confession", and they then quote these words of Hermann Weyl[36]:

♣ Have any of my readers seen anything in Whitehead's writings to support Adams' contention?

[36] in *Mathematics and Logic*, a brief survey serving as preface to a review of "The philosophy of Bertrand Russell", *Am. Math. Monthly* **53** (1946) 2–13.

"We are less certain than ever about the ultimate foundations of (logic and) mathematics. Like everybody and everything in the world today, we have our "crisis". We have had it for nearly fifty years. Outwardly it does not seem to hamper our daily work, and yet I for one confess that it has had a considerable practical influence on my mathematical life: it directed my interests to fields I considered relatively "safe", and has been a constant drain on the enthusiasm and determination with which I pursued my research work."

Weyl's statement rebuts Dieudonné's suggestion that foundational work has had no influence: Weyl, for one, was influenced to move away from areas where the paradoxes had manifested themselves.

So I am not convinced by Dieudonné's attempt to link logic and set theory and separate them from the rest of mathematics; each mathematician sets his own boundaries; there are among my acquaintanceship mathematicians with whom I have fruitful exchanges when I am in set-theorist mode, but whose eyes glaze over should I ever say anything to them when I am in logician mode.

D·11 In the section on *les conceptions de Bourbaki*, he says that Bourbaki have invented nothing but have restricted themselves to making explicit the practice of those mathematicians called "formalists", who, as we shall see, (or rather, as Dieudonné will tell us) form the quasi-totality of mathematicians of today.

II.7: Bourbaki has only two things to say: *all are free to think what they will about the nature of mathematical entities or about the truth of the theorems they use, so long as their proofs can be transcribed into the common language*

D·12 ASIDE — the common language being that imposed by Bourbaki —

and as for contradictions, Bourbaki *believes in doing nothing till an actual contradiction occurs.*

D·13 COMMENT Really what he is saying is that most mathematicians are content to be ignorant, and Bourbaki wishes to reinforce that ignorance.

Formalists are interested in objects ... the interpretations of a system of signs that is subject to a rigorous syntax that is independent of any interpretation.

D·14 COMMENT I suspect Dieudonné thinks, reasonably enough, that all minds are different and that the common ground is to be subscription to a particular syntax.

In the syntax are the rules of classical logic and axioms of ZF.

D·15 COMMENT That statement is false, for Bourbaki omit the axiom of foundation, which is not derivable from the axioms that they give: the issue, for Logicians, here is that of collection versus replacement.

Dieudonné shirks the question of why these particular rules should be adopted; the "common language" appears to be one imposed by fiat, of which no discussion is permitted, and hence the intuition is paralysed.

II.3: *ZF gives only the bare theory of sets, which 95% at least of today's Mathematicians consider to be without interest.*

D·16 COMMENT I wonder what Dieudonné means by "the bare theory of sets". It might be that Dieudonné is confusing two things: the theory of sets and the theory of **Set**, the category of sets, of which the objects are indeed sets without any structure, about which, taken in isolation, there would be little to say. But post-Zermelo set theory studies the membership relation $\in$, and that relation, when the axioms of ZF are assumed, provides so rich a universe that large amounts of mathematics can be expressed within it.

II.4: *the axioms of a structure are disguised definitions.*

D·17 COMMENT One feels that for Dieudonné a structure is a type rather than a token.

D·18 Bourbaki evidently believe they have provided a formalism that is adequate for the quasi-totality of Mathematicians. They also seem very keen that no-one should examine the formalism that they provide. They have not explained why their chosen axiomatisation—somewhere between ZFC without foundation and ZF with global choice—is to be the standard. Indeed, given that Bourbaki, as Corry [Cor1, 2] has shown, ignore their own foundations, why should others be expected to use them?

II.6: *the quasi totality of Mathematicians are naïf pre-Cantorians.*

D·19 COMMENT Is that desirable? Grothendieck in his re-structuring of algebraic geometry needed some set theory, but he only knew the cruder parts of set theory as presented in Bourbaki's book: how might his work have advanced if he had had the subtler parts of set theory at his fingertips?

D·20 COMMENT As for the "pragmatic" suggestion that no one should think about possible contradictions: in one way it is sensible; in another it is idiotic. Dieudonné has used CH: was Gödel wasting his time in seeking and finding a consistency proof for it? I am not saying that people *should* go out of their way to think about possible contradictions; but, as Saccheri found, looking for a consistency proof is much the same as looking for an inconsistency proof; and sometimes one finds a "proof" of a contradiction which turns out to be a proof of something quite different.

D·21 In the résumé Dieudonné says that Bourbaki's system is implicit in the work of the quasi-totality of mathematicians; and they are all happy to be naïve about philosophical questions. Is that *really* so desirable?

Dieudonné in effect is saying in a patronising manner,

"NO NEED FOR ANY OF YOU YOUNG CHAPS TO WORRY YOUR HEADS ABOUT FOUNDATIONS, WE HAVE DONE IT ALL FOR YOU."

That is a dangerous position, for it is an attempt to gag the future. Look how Gordan criticised Hilbert for his refusal to present a certain proof within the then accepted rules; or turn to the book [Ad] of J. Frank Adams. On page 293, in section 14, "A category of fractions", he writes:

> "*(Added later.) I owe to A.K.Bousfield the remark that the procedure below involves very serious set-theoretical difficulties. Therefore it will be best to interpret this section not as a set of theorems, but as a programme, that is, as a guide to what one might wish to prove.*"

and on page 295, presumably discussing the same difficulty, he writes:

> "*(Added later: unfortunately there is no reason why the result should be a small category.)*"

In other words, he has essayed a localisation construction, and it founders on his inadequate grasp of set theory. Later Bousfield [Bous1,2] found a rigorous, set-theoretically correct replacement for Adams' ill-founded argument. and very recently Fiedorowicz [AF] has shown how to correct Adams' original lectures.

D·22 The point is enriched by a paper of Casacuberta, Scevenels and Smith [CaScS] entitled "Implications of large-cardinal principles in homotopical localization." They find that the question "Is every homotopy idempotent functor equivalent to localization with respect to some single map?", which was motivated by Bousfield's discovery, cannot be answered in ZFC, for they find that their localisation principle actually implies the existence of fairly large cardinals, and follows from the existence of the even stronger hypothesis known as Vopěnka's principle; thus their localisation principle inescapably involves those same large cardinals that Dieudonné was so confident[♠] the quasi-totality of mathematicians would never encounter.

More recently [BCM, BCMR], Bagaria, Casacuberta, Rosický and the present author have obtained similar results from a weaker hypothesis than Vopěnka's principle, namely the existence of a supercompact cardinal. Casacuberta and Rosický think in terms of categories, whereas Bagaria and I are set theorists, and we have had to keep translating from one perspective

---

[♠] as was Mac Lane.

to the other, each perspective contributing to the outcome; a set-theoretical argument having to be re-worded so that it might have meaning in categories where the objects do not have members; an example from category theory having to be explained to set theorists. The experience has only reinforced my belief in the necessity of a pluralist account of the foundations of mathematics.

D·23 To look briefly at the other non-mathematical essays in [Di2]: in his piece on *L'évolution de la pensée mathématique dans la Grèce ancienne*, Dieudonné argues against Launcelot Hogben's view that mathematics is utilitarian. Thus he attacks logic using the same weapons that others use to attack mathematics, attacks he would wish to repulse.

In his essay, *Liberté et Science Moderne,* he writes: *"Il faut, pour pouvoir faire des découvertes en science, avoir l'audace de contredire les idées reçues"* — a bit rich, that, but not as rich as his final paragraph on the philosophy of Bourbaki:

> *"La plus charitable hypothèse est de penser que cela n'est dû qu'à l'ignorance, ou au refus de s'informer, ou à l'incompréhension; sinon, il faudrait conclure qu'il s'agit d'illuminés aveuglés par leur fanatisme, et que la "crise" qu'ils croient voir dans les mathématiques d'aujourd'hui ne se trouve que dans leur cerveau."*

In *L'abstraction et l'intuition mathématique* he talks sense: he says there is more than one intuition—precisely the grounds on which I argue for a pluralist view of the foundations of mathematics—and that there are transfers between intuitions; here and in *Liberté* he is arguing in favour of those conditions that favour creativity; precisely those conditions which, if claimed by logicians, he so strongly rejects.

D·24 It should perhaps be remarked that in the first volume *Foundations of modern Analysis* [Di1] of his treatise, Dieudonné, though referring the reader to [Bou54] for a formal axiomatisation, lists in his Chapter I axioms of a system that is, one ambiguity aside, essentially Bou49 and not Bou54: the axioms he mentions are extensionality, singleton, power set, the scheme of separation, ordered pairs, cartesian products, and the axiom of choice. He discusses families of subsets of a given set; the ambiguity is that when, later, he enunciates the principle that the union of a countable family of countable sets is countable he does not state that the elements of the family are required to be subsets of a given set, though he might have intended that restriction, since he offers a proof. But the proof reads slightly oddly, since it uses, as it must, the axiom of choice, but does not mention it. On the other hand Dieudonné quite unnecessarily states as an additional axiom the easy consequence of the axiom of choice, that each infinite set has a denumerably infinite subset.

D·25 In his last book, [Di3], Dieudonné makes the same mistake that he made in his position papers of fifty years previously: he went to his grave believing that truth and provability are identical.

E: ... and fostered by the errors and obscurities of a well-known undergraduate textbook,

TURN NOW TO A TEXTBOOK that has had a wide following in France: *Tome 1, Algèbre,* of the *Cours de mathématiques* by Jacqueline Lelong-Ferrand and Jean-Marie Arnaudiès, anciens élèves de l'École normale supérieure. This book is the first of a four-volume treatise of mathematics, which is described by Alain Pajor as a classic which students often consider to be difficult and use like a dictionary. Its first edition was in 1978; I work from the third edition, of 1995. Ominously, among the works cited in its *Bibliographie* are these two:

N. Bourbaki, *Théorie des ensembles,* published by Hermann in 1957;

R. Godement, *Cours d'algèbre,* Hermann, edition of 1966;

and the influence of those two works reveals itself in the choice and the bias of the presentation of logic and set theory in Chapter I of this textbook, which chapter, though, displays a further degeneration of clarity, correctness and coherence as compared with the above two sources. Indeed I propose to criticise Chapter I rather severely. In doing so I aim to show, first, that what the authors have to say on foundational themes is often lamentably vague and, worse, in places actually false; and, secondly, that, some of their Delphic pronouncements being only interpretable in the light of their sources, their account must be seen as deriving from those of Hilbert, Bourbaki and Godement, and thus as continuing the descent into incoherence; from this perspective the banning of logic from the CAPES might be a reasonable consequence of the belief that seems to have taken hold, that the subject is too messy to inflict on the young and on their teachers.

My typographical conventions in this section: passages in *slanted type,* if in French, are taken direct from their text; if in English, are my translation of a passage from the text. Passages in roman type are my commentary on their text, with **bold face** used to emphasize certain of my comments. I use *italic type* to highlight certain phrases, most often reproducing a highlight in their text. A centred section heading **[L-F,A] I.2** marks the beginning of my main criticism of their Chapter I, section 2. A marginal reference [LFA] p2,-6 will be to page two of their book, line six from the bottom.

The first four divisions of [L-F, A] Chapter I are labelled

1. *Notions sur la formalisation*
2. *Règles de logique formelle*
3. *Quantificateurs*
4. *Opérations sur les ensembles*

and contain respectively

> 1. remarks on the need for formalisation, and some comments on syntax
> 2. remarks on and some axioms for propositional logic
> 3. remarks on the meaning of ∃ and on the Axiom of Choice
> 4. introduction of certain set-theoretic axioms.

Each of those four requires criticism; in the rest of the chapter, as the authors get closer to areas of mathematics with which they are familiar, the need for comment diminishes but a few remarks will still be necessary.

The order in which topics are treated is sometimes unexpected; there is an early mention, on page 2, of the need for an axiom asserting the existence of the set of natural numbers; on the other hand a definition of "finite" is not given till page 32, though at some language level the concept is being used on many previous pages, such as 10, 11, 12, 17, 20, 21.

The bulk of the discussion of the Axiom of Choice takes place before most of the axioms of set theory have been presented; this ordering of topics might be the result of the implicit reliance on the epsilon operator and its consequent representation of the Axiom of Choice as a principle of logic rather than of set theory.

[L-F,A] I.1: syntax

The authors begin by commenting briefly that certain paradoxes can be avoided by constructing formalised languages, which are less expressive than natural languages, and then developing formal logics. Thus the theory of sets may be built up by the axiomatic method:

[LFA] *On se donne un petit nombre de*	One gives oneself a small number
p 1 *signes logiques, et un petit nombre*	of logical signs, and a small num-
de règles permettant, à l'aide de	ber of rules permitting with the
ces signes, et des lettres des divers	aid of these signs and letters of
alphabets, d'écrire des "mots per-	various alphabets, the writing of
mis". (Le plus souvent, les mots	"permitted words". (The permit-
permis s'appellent des assemblages.)	ted words are most often called
	assemblies.)

E·1 COMMENT Although they do not set out the rules of formation of languages, I suspect they intend only formulæ with quantified set variables to be considered.

[LFA] *On se donne un moyen de distinguer* | *Among the permitted words or as-*
p 1 *deux sortes de mots permis: les uns,* | *semblies there are terms, (which*
appelés termes, seront les répresen- | *are abstract representations of the*
tants abstraits des objets sur lesquels | *objects about which one will rea-*
on fera des raisonnements; les autres, | *son) and there are relations which*
appelés relations représenteront les | *represent the assertions that one*
assertions que l'on peut faire sur | *can make about these objects. ..*
ces objets. Puis on se donne des | *then one gives oneself some rules*
règles régissant l'usage des relations, | *governing the usage of relations,*
permettant de construire de nou- | *permitting the construction of new*
velles relations à partir de relations | *from old; these are the rules of for-*
données, &c; ces règles sont les règles | *mal logic.*
de logique formelle.

E·2 COMMENT We see that whereas Bourbaki held to a strictly formalist line that their system was an uninterpreted calculus, the present authors allow a difference between terms and the objects that are their interpretation. However they give no rules for deciding which assemblies are terms.

[LFA] *Cela étant fait, la notion de verité* | *The notion of mathematical truth*
p 1 *mathématique est "relativisée" de* | *is "relativised" in the following*
la manière suivante: on pose un | *manner: a small number of rela-*
petit nombre de relations, appelées | *tions, called axioms, are supposed*
axiomes, comme vraies a priori. Puis | *true a priori. Then one defines*
on définit la notion de démonstration. | *the concept of a formal proof.*

E·3 COMMENT As with defining the notion of "term", the authors leave the notion of a formal proof all too vague. It is not clear what are the axioms nor what are the rules of inference.

Two most revealing remarks are these:

[LFA] *une relation est alors dite vraie si* | *A relation is called true if it may*
p 2 *elle peut être insérée dans une* | *be inserted in a proof.*
démonstration.

[LFA] *une relation est vraie si on peut l'in-* | *A relation is true if one may in-*
p 3 *sérer dans un texte démonstratif.* | *sert it in a demonstrative text.*

E·4 COMMENT This equation of truth and provability is, of course, standard for horses from the Bourbachiste stable. But it is far from satisfactory.

There has been a suggestion that any relation might be taken to be an axiom, and therefore might be used in a proof. For example, both the Continuum Hypothesis, CH, and its negation have interesting consequences. So either of them might be used in a proof, and therefore, in this odd meaning of "true", both are true.

Thus the notion of "true" is imprecise; mutually contradictory systems certainly exist.

On page 5, they speak of *axiomes momentanés*, thus acknowledging that one might wish to postulate hypotheses for an argument, to be discharged once the argument is complete.

[LFA] *Lorsqu'on dispose d'un langage for-* p2, 6 *malisé cohérent pour fonder une théorie (la théorie des ensembles, ou toute autre théorie mathématique), le développement de cette théorie consiste à en trouver les relations vraies, auxquelles on donne le nom de théorèmes, propositions, lemmes, scholies etc.*

Once one has a formalised language available for basing a theory (such as the theory of sets, or any other mathematical theory) the development of that theory consists of finding true relations which one calls theorems, propositions, lemmata, scholia, &c.

[LFA] *Signalons qu'on peut construire tou-* p2,-6 *tes les mathématiques connues à ce jour à l'aide des axiomes et du langage formalisé de la théorie des ensembles; l'axiome fondamental étant l'existence d'au moins un ensemble de nature mathématique: celui des nombres entiers.*

Let us note that one can construct all the mathematics known today with the aid of the axioms and formalised language of the theory of sets; the fundamental axiom (for mathematicians, at least) is the assertion of the existence of at least one set of a mathematical kind, namely the set of whole numbers.

[LFA] *Mais en pratique, il est impossible* p 3 *d'écrire toutes les mathématiques en langage formalisé: pour le moindre théorème facile, il faudrait des livres entiers. On est donc amené à utiliser des abréviations et du langage courant; et on se contente d'écrire des textes "dont on est sûr" qu'on pourrait les formaliser.*

But in practice it is impossible to write mathematics entirely in a formal language; for even the easiest theorem it would need whole books. So one contents oneself with writing texts of which one may be sure that they could be formalised.

E·5 COMMENT The claim made in the second of those three paragraphs echoes that at the end of Bourbaki's address [Bou49] and is shown in [M10] to be erroneous. We shall also see that, curiously, the authors do not give the existence of **N**, the set of natural numbers, as an axiom, but state that its existence can be derived from a less precisely phrased assertion of the existence of an infinite set.

E·6 COMMENT The third paragraph, of course, echoes the remarks of Bourbaki and Godement about the length of expansions of the formulæ of their chosen language.

[L-F,A] I.2: propositional logic

[LFA] *Étant donnée une relation A, on*	*Given a relation A, one defines its*
p 3 *définit son contraire, notée non A;*	*contrary, written (not A); (not A)*
(non A) est aussi appelée la négation	*is also called the negation of A.*
de A. Par définition, A est fausse si	*By definition, A is false if not-A*
(non A) est vraie.	*is true.*

E·7 COMMENT What does "false" mean? Perhaps that is a definition. But how to know if not-A can be used in a proof?

[LFA] *S'il existe A telle que A et (non A)*	*If for some A, both A and not-A*
p 3 *soient vraies, la théorie est dite con-*	*are true the theory is called con-*
tradictoire et on démontre qu'alors	*tradictory, and then every relation*
toute relation de la théorie est vraie.	*of the theory may be proved to be*
Bien que cela n'ait pas été démontré,	*true. It is generally supposed that*
on pense généralement que la théorie	*the theory of sets is not contradic-*
des ensembles n'est pas contradic-	*tory, although that has not been*
toire, de sorte que pour toute re-	*demonstrated, so that for each re-*
lation A de cette théorie, l'une au	*lation of this theory at most one*
plus des relations A et (non A) est	*of the relations A and (not A) is*
vraie.	*true.*

E·8 COMMENT As often for writers of the Bourbachiste persuasion, Gödel is a non-person and no mention is made of the reason why the consistency of set theory has not been demonstrated.

They go on to mention the possibility that there are some statements that are neither provable nor refutable; and seek to marginalise this shocking phenomenon in the standard Bourbachiste way:

[LFA] *Il ne faut pas croire que l'une des*
p 3   *deux relations A et (non A) soit*
       forcément vraie: il pourrait exis-
       tait en effet des relations contraires
       A et (non A) telles qu'aucune des
       deux ne puisse être insérée dans une
       texte démonstratif; A est alors dite
       indécidable. En pratique nous ne
       rencontrerons pas de relation indé-
       cidable !

It should not be thought that nec-
essarily one of the relations A and
(not A) is true: there might exist
in fact contrary relations A and
(not A) such that neither of them
can be inserted in a demonstrative
text; A is then said to be undecid-
able. In practice we do not meet
them.

E·9 COMMENT That was Godement's original hope and also Dieudonné's, except of course when he found he could use the Continuum Hypothesis to construct a counterexample to something. But whereas Godement by 1969 was acknowledging the existence of undecidable statements of ordinary mathematics, these followers of his were still denying that existence in 1995.

[LFA] *Étant données deux relations*
p 3   *A et B on définit la disjonc-*
       tion de A et B, notée: (A ou
       B). Si l'une au moins des re-
       lations A, B est vraie, (A ou
       B) est vraie.

Given two relations A and B, one
defines their disjunction, written
A-or-B. If at least one of A or B
is true then (A-or-B) is true.

E·10 COMMENT That seems much too vague; how might one say that A-or-B is the strongest such statement? Indeed below, they remark that (A or (not A)) is always true, whereas above they have admitted that for certain A, neither A nor (not A) is true.

[LFA] *La relation ((non A) ou B) s'appelle*
p 3   *l'implication de B par A, et se note*

$$A \implies B.$$

Si A et $(A \implies B)$ sont vraies, B
est vraie; si B est vraie, $(A \implies B)$
est vraie pour toute relation A.

The relation (not-A or B) is called
the implication of B by A, and is
written

$$A \implies B.$$

If A and $A \implies B$ are true, so is
B; if B is true, $A \implies B$ is true
for every A.

E·11 COMMENT Of those two comments, the first is *Modus Ponens*, the rule of inference given in Bourbaki and used in *Principia Mathematica*. The second is a derived rule, given as C9 in Bourbaki.

[LFA]
p 3
De plus, nous admettrons les relations ci-dessous, qui fournissent des règles de raisonnement:

Further, we shall admit the following relations, which supply the rules of reasoning:

1) $A \implies A$. *Cette relation est très interessante, car elle exprime que $(A$ ou $($non $A))$ est toujours vraie, (même si A est undécidable !)*

1) $A \implies A$. This relation is very interesting, for it says that $(A$ or $($not $A))$ is always true (even if A is undecidable !)

2) $(A$ ou $A) \implies A$.
3) $A \implies (A$ ou $B)$.
4) $(A$ ou $B) \implies (B$ ou $A)$
5) $(A \implies B) \implies [(C$ ou $A) \implies C$ ou $B]$.
6) $(A \implies B) \implies [(B \Rightarrow C) \implies (A \Rightarrow C)]$.
7) $(A \implies ($non$($non$A)))$.
8) $[A \implies B] \implies [($non $B) \implies ($non $A)]$

2) $(A$ or $A) \implies A$.
3) $A \implies (A$ or $B)$.
4) $(A$ or $B) \implies (B$ or $A)$
5) $(A \implies B) \implies [(C$ or $A) \implies C$ or $B]$.
6) $(A \implies B) \implies [(B \implies C) \implies (A \Rightarrow C)]$.
7) $(A \implies ($not$($not$A)))$.
8) $[A \implies B] \implies [($not $B) \implies ($not $A)]$

Parmi les règles ci-dessus, les règles numéros 1), 2), 3), 4), 5) sont des axiomes dans la plupart des logiques formelles usuelles.

Of the above rules, numbers 1) to 5) are axioms in most customary logical systems.

E·12 COMMENT Of those axioms, 2) to 5) are exactly S1, S2, S3, S4 of Bourbaki and essentially *1.2, *1.3, *1.4 and *1.6 of *Principia Mathematica* and AL1, AL2, AL3 and AL4 of Godement. The others are consequences of them, 1), 6), 7), 8) being respectively C8, C6, C11, C12 of Bourbaki and TL2, TL1, and essentially TL3, and TL4 of Godement.

[LFA]
p 4
Si A et B sont de relations, on définit la conjonction de A et B, notée $(A$ et $B)$: c'est la relation:

If A and B are relations, their conjunction, in symbols $(A$ et $B)$, is defined as the relation

non $[($non $A)$ ou $($non $B)]$.

not $[($not $A)$ or $($not $B)]$.

On dit que A et B sont équivalentes si $(A \implies B)$ et $(B \implies A)$ sont vraies, on écrit alors $(A \iff B)$.

One says that A and B are equivalent if the two implications are true. One then writes $A \iff B$.

E·13 COMMENT As far as I can tell, that implies that $(A \Longleftrightarrow B) \Longleftrightarrow C$ is meaningless. According to the text, the sequence of symbols "$A \Longleftrightarrow B$" expresses the conjunction of the two statements

$\qquad$ "$A \implies B$" *can be inserted in a proof*

and

$\qquad$ "$B \implies A$" *can be inserted in a proof*

and is thus not a formula but a statement about two formulæ and a theory. But if Φ is a statement about a system and C is a formula of the system, neither $C \implies \Phi$ nor $\Phi \implies C$ can be inserted in a proof since neither is a formula of the system.

[L-F,A] I.3: predicate logic and the Axiom of Choice

[LFA] *Nous écrirons*
p 6

$\qquad$ (1) $\qquad \exists x, \quad A(x)$

pour exprimer la relation "la relation $A(x)$ est vraie pour au moins un objet x". Cette définition n'est qu'intuitive, nous ne ferons que décrire les règles d'usage du symbole $\exists$, appelé quantificateur existentiel.

We write

$\qquad$ (1) $\qquad \exists x, \quad A(x)$

to express the relation "the relation $A(x)$ is true for at least one object x". This definition is only intuitive, we only make it to describe the rules of use of the symbol $\exists$, which is called the existential quantifier.

E·14 COMMENT Here we have a confusion of language levels. Relations have been defined as certain assemblies, that is certain strings of formal symbols. The statement within quotation marks is not a relation; it is a (French) phrase about relations and objects. So the authors have achieved precisely that confusion of language levels (as exploited by the paradoxes) that the introduction of formalised languages was intended to avoid.

[LFA] *Nous écrirons*
p 6

$$\forall x, A(x)$$

pour exprimer la relation

(2) $non(\exists x, nonA(x))$

Cette relation signifie que la propriété $A(x)$ est vraie de tous les objets x.

We write

$$\forall x, \ A(x)$$

to express the relation

(2) $not\text{-}(\exists x, \ not\text{-}A(x))$

This relation signifies that the relation $A(x)$ is true for all objects x.

E·15 COMMENT We verge on a problem of ω-incompleteness here. In a theory of arithmetic it could easily be the case that each $A(\mathfrak{n})$ is provable but that $\forall nA(n)$ is not; if objects are the same as terms and truth is provability, then "the relation $A(x)$ is true for all objects x" says that each $A(\mathfrak{n})$ is provable, but that does not mean that $\forall nA(n)$ is provable.

E·16 REMARK The authors state that the formulæ (1) and (2) and the rules of logic permit the mechanical use of quantifiers. But what are those rules? They are not stated, though some examples are given.

page 7: the problem of choice

We come now to the six paragraphs that Lelong-Ferrand and Arnaudiès devote to commenting on the nature of the Axiom of Choice. Here I use the sign ¶ to mark the start of my discussion of one of the six.

¶1 The first paragraph mentions the problem of the meaning of an existential statement: some great mathematicians such as Émile Borel have not believed non-constructive proofs of an existential statement.

[LFA] *Intuitivement, le problème se pré-*
p 7 *sente comme il suit: peut-on démontrer, dans une théorie donnée, un théorème de la forme:*
$(\exists x, A(x))$ sans construire, par un procédé descriptif, un object x pour lequel la relation $A(x)$ est effectivement vraie?

Intuitively the problem is this: can one prove, in a given theory, a theorem of the form $(\exists x, A(x))$ without constructing, by a descriptive procedure, an object x for which the relation $A(x)$ is actually true?

E·17 COMMENT Those with a taste for constructive proofs usually eschew the Axiom of Choice.

¶2   The second, dreadfully confused, and actually wrong (rather than stylistically undesirable or misleading) paragraph alleges that the axiom says that when an existential formula is true one can always formally construct a witness:

[LFA] *On a vite reconnu la nécessité d'in-*
 p 7   *troduire, en théorie des ensembles,*
     un axiome appellé axiome du choix:
     grosso modo, cet axiome dit que
     lorsqu'une relation du type $\exists x,$
     $A(x)$, *est vraie, on peut toujours*
     construire formellement un objet x
     pour lequel $A(x)$ *est vraie.*
	The necessity of introducing an ax-
	 iom called axiom of choice into the
	 theory of sets was quickly recog-
	 nised; roughly, the axiom says
	 that when a relation of the type
	 $\exists x, A(x)$ *is true, one can always*
	 formally construct an object x *for*
	 which $A(x)$ *is true.*

E·18   COMMENT **That is false:** the Axiom of Choice implies, for example, that there is a well-ordering of the continuum, but that is perfectly compatible with there being no definable such.

One wonders if this mistake is related to one noted by Alonzo Church in his 1948 review [Chu] of "L'énumeration transfinie. Livre I. La notion de rang", by Arnaud Denjoy [De]:

> "in an otherwise excellent work, the treatment (pp 5, 110-116) of the Axiom of Choice and of Zermelo's theorem that every class can be well-ordered, is without value, because the author mistakenly identifies the Axiom of Choice with the proposition that every non-empty class has a unit subclass."

¶3   The third paragraph says that this axiom has many equivalent formulations, of which the best known are Zermelo's axiom concerning well-ordered sets, and the theorem of Zorn.

E·19   COMMENT True, in that the Axiom of Choice, as usually understood (but not as presented by our authors) is indeed equivalent, as proved by Zermelo, to the proposition that every set can be well-ordered, and as proved by Zorn, to the proposition known in Anglophone countries as Zorn's Lemma.

¶4   The fourth paragraph is highly revealing:

[LFA] p 7 *Dans la mathématique formelle usuelle, l'axiome de choix est introduit dès le départ, à l'aide d'un signe logique. Dans la théorie des ensembles ainsi construite, le symbole $\exists x, A(x)$ n'est qu'une abréviation pour exprimer, en quelque sorte, que l'objet théorique qu'il est possible de construire et qui vérifie $A(x)$, vérifie effectivement cette relation.*	*In mathematics as usually formalised, the Axiom of Choice is introduced at the start by the aid of a logical sign. In this presentation of the theory of sets, the symbol $\exists x, A(x)$ is only an abbreviation for expressing in some manner that the theoretical object that it is possible to construct and which satisfies $A(x)$ does indeed satisfy this relation.*

E·20 COMMENT I have no idea how to interpret their remarks; but I can say where they came from, namely the use of the Hilbert ε-operator, the one called τ by Bourbaki.

¶5   The fifth paragraph remarks correctly that set theories without the Axiom of Choice have been studied, and that one can therefore classify results according to their dependence or otherwise on that axiom.

¶6   The sixth and last paragraph on page 7 lists some example of existence statements that require the Axiom of Choice for their proof; their first example is erroneous, though the others are correct. The statement that the Axiom of Choice is needed to prove the existence of an arbitrary product is inaccurate; AC is needed to prove that the product is non-empty if the factors are, not that the product exists.

Fortunately, at the top of page 21, they state correctly that:

Si, pour tout $i \in I, A_i \neq \varnothing$, $$\text{alors } \prod_{i \in I} A_i \neq \varnothing.$$ *Cette propriété est un axiome équivalent à l'axiome du choix.*	*If, for each $i \in I, A_i \neq \varnothing$,* $$\text{then } \prod_{i \in I} A_i \neq \varnothing.$$ *This property is an axiom equivalent to the Axiom of Choice.*

[L-F,A] I.4: operations on sets

The authors now discuss various operations on sets, and mention various justificatory axioms. But they make conflicting statements: at the top of page 8 they say that a set is a term equipped with a relation $\in$. Lower down, they say that it is hard to tell which terms are sets.

[LFA] *Nous admettons la notion d'en-* p8, 1 *semble. Un ensemble est donc un terme, muni d'une relation:* ∈.	*We accept the notion of a set. A set is then a term, equipped with a relation:* ∈.
— *La relation :* $a \in E$ *se lit* " *a appartient à E* "	— *The relation :* $a \in E$ *is read* " *a belongs to E* "
⋮	⋮
[LFA] *On n'a aucun moyen effectif de re-* p8,-11 *connaître si un terme donné est un ensemble; aussi la théorie des ensembles procède par construction, à partir de termes dont on admet qu'ils sont des ensembles (par exemple,* **N**).	*We have no effective means of telling whether a given term is a set; so the theory of sets proceeds by construction, starting from those terms that are acknowledged to be sets, (for example* **N**).

E·21 COMMENT They are, I suspect, in the condition of the patient in Laing's double-bind model: they have been given contradictory statements by people they regard as authorities. On the one hand, Zermelo held that sets are those classes which are small enough to be members of some class, whereas proper classes are those classes which are too big to be a member of any class. So if one has a class $\{x \mid R\}$, to say that it is a set is to say that $\exists y \; y = \{x \mid R\}$; and there are certain classes, such as the Russell class $\{x \mid x \notin x\}$ of which the set-hood is refutable. On the other hand, we saw in Section B that Bourbaki's use of the symbol $\{x \mid R\}$ is not the same as Zermelo's. With Bourbaki, each term is, by syntactical trickery, provably equal to some set; thus

$$\vdash_{\mathsf{Bou54}} \exists y \; y = \{x \mid x \notin x\} \quad \text{whereas} \quad \vdash_{\mathsf{ZF}} \neg \exists y \; y = \{x \mid x \notin x\}.$$

So at the top of page 8, the authors are with Bourbaki, but lower down they are with Zermelo.

The translation into Bourbaki's dialect of set theory of the assertion in Zermelo's dialect that the class $\{x \mid R\}$ is a set, is the formula $Coll_x(R)$. Reassuringly,

$$\vdash_{\mathsf{Bou54}} \neg Coll_x(x \notin x).$$

E·22 COMMENT In fact the authors are cautious with the use they make of the $\{ \cdot \mid \ldots \}$ notation. They introduce at the top of page 9 the notation $\{x \mid x \in E \text{ et } A(x)\}$, for use only when E is a set and $A(x)$ is a relation, to mean the set of those elements of the set E which have the property A;

and on page 8, they use $\{a, b, c, \ldots, l, m\}$ as a notation for (presumably) a finite set, though they do not explain their use of three dots to the reader.

E·23 COMMENT The use of the word "relation", as used at the top of page 8, is abnormal: for example, a total ordering is a set together with a relation that might hold between two members of that set, but here the relation $\in$ is one that holds between a member of the set and the set itself. The same letter $\in$ is used for the relation associated to any set. Why, I wonder?

Then in the third line of text on page 8, "relation" is used in the sense introduced on page 1, to mean a permitted word that is interpretable as an assertion.

On page 1 it is stated that a term is a permitted word that is an abstract representation of one of the objects about which we wish to reason; I do not see how a word, a finite string of symbols, can be *muni d'une relation:* $\in$; though I can believe that the entity abstractly represented by that word might be.

Discussion of equality

At the top of page 8, $a = b$ is informally introduced: no axioms for equality have hitherto been given, though the sign $=$ occurs in examples on page 5.

The authors say that $a = b$ expresses that a and b are the same object, and that intuitively (page 8 again) a set E is the collection of objects that are members of E.

On page 8: unordered pairs, triplets, etc are stated to exist, though no axiomatic justification is given or claimed for that. Three dots are used, to suggest a finite but arbitrarily long sequence; the word "finite" is not used here.

On page 8: what set theorists know as the axiom of extensionality, that two sets with the same members are equal, is formulated but not stated to be an axiom.

E·24 COMMENT On page 9, Theorem I.4.1 is stated and "proved", that the empty set exists and is unique. To prove its uniqueness some form of extensionality would normally be required. To prove its existence, the existence of some set must be asserted; and so far no axiom says that, so I suppose that the authors are tacitly relying on the fact that it is a theorem of Hilbertian logic with equality that something exists.

A list of axioms of set theory

Now the authors proceed to state various axioms of set theory, but there is nothing to indicate when their listing of axioms has reached its end.

AXIOM 1   On page 8, near the bottom, the separation scheme is given.

AXIOM 2   On page 9, line -7, the power set axiom is given.

AXIOM 3   At the top of page 10, the existence of the ordered pair, *le couple*, (a, b), of two terms a and b, is stated to be an axiom; presumably the principle stated at the bottom of page 9,

$$((a', b') = (a, b)) \Longleftrightarrow ((a = a') \ \& \ (b = b')),$$

should be included in this axiom.

AXIOM 4   Page 10, line 6, cartesian products: the existence of $E \times F$ is an axiom.

E·25   ASIDE   Cartesian product is then stated to exist for **any finite number** of terms, though nothing is said about associativity.

E·26   REMARK   The existence of the intersection of two sets is, correctly, derived from the scheme of separation; the existence of the union of two sets is said to follow from an (unstated) axiom. The principal construction of [M10] shows that the statement, that if x and y are sets then so is $x \cup y$, does not follow from the axioms that the authors actually formulate.

E·27   REMARK   The axioms are stated to imply the existence of finite sets; whence, the authors say, one can define the integers and develop some number theory.

AXIOM 5   At the top of page 12, it is said that the existence of the set **N** of whole numbers requires a new axiom, called the axiom of infinity, which states that "there is a set which is not finite".

E·28   REMARK   The authors state that once one has **N**, one can construct all the sets used in usual mathematics.

   No definition is given at this point of "finite", nor is *entier* defined; nor is any derivation offered from the stated axiom of infinity that the set of whole numbers exists,

   No further axioms are listed: Thus their system is essentially Bou49, though presented with less precision.

The remaining divisions of [L-F,A] Chapter I

E·29   On page 17, the authors begin their discussion of indexed families of sets, but their families are always taken to be always subsets of some given set. Thus given two sets which are subsets of the same set X say, their union, being a subclass of the set X, can be proved to be a set by separation; but, as mentioned in Remark E·26, their system fails to prove that the union of two arbitrary sets is a set. Bou49 is thus weaker than the system Bou54 used in their two cited sources [Bou54] and Godement [Gd], who all admit indexed families of sets which are not necessarily subsets of a set given in advance, thus obtaining a system that includes a form of the axiom of replacement; whereas Bou49 is essentially the system of Zermelo without foundation, without unordered pairing, but with ordered pairs, cartesian products and a global form of choice.

E·30 No definition of *finite* is given before page 32: as with Godement the definition is very late in arrival, but is used earlier. Some of those earlier uses are really of finiteness in the metalanguage; but no such distinction is made by the authors.

N is mentioned on page 8 as an object that will be admitted to be a set. **N** is used on page 20, when sequences are introduced, and again on page 28, to provide an example of an ordered set; it is emphasized that that ordering is a well-ordering. Three pages later, our hopes of a definition of **N** are dashed:

[LFA] *Nous supposons connues toutes les* p 31 *définitions et propriétés élémentaires relatives aux nombres entiers naturels. Tout au long de l'ouvrage, l'ensemble des nombres entiers sera désigné par* **N**.	*We suppose known all the definitions and elementary properties relating to the natural numbers. Throughout the work, the set of (non-negative) integers will be denoted by* **N**.

E·31 Finally, on page 32, we reach the long-awaited definition: a set is said to be *finite* if it is in bijection with an initial segment of the natural numbers.

E·32 COMMENT **That is circular.** The axiom of infinity was formulated as "There is a set which is not finite", but without a definition of *finite* having been given. We were told that one can derive the existence of **N** from the existence of a set which is not finite, again without a definition of *finite*. Now we are told that a set is finite if it is in bijection with an initial segment of the natural numbers.

If one tries to interpret those statements in a way that removes the circularity, one arrives at the statement that there is a linearly ordered set which is not in bijection with any proper initial segment of itself. But that is insufficient: the set $\{0, 1, 2, 3, 4\}$ has a linear ordering under which it is not in bijection with any proper initial segment of itself; but no one would say that set was infinite.

E·33 REMARK Godement avoids this trap: he uses Dedekind's definition of *finite* and then defines **N** as the set of finite cardinals. But our authors, on page 36, speak only of infinite cardinals.

On page 32, it is stated that if A is not finite there is an injection of **N** into A: a covert use of AC.

Discussing the continuum hypothesis, Lelong-Ferrand and Arnaudiès write:

[LFA] *Depuis 1966 (travaux de l'améri-* p 37 *cain Cohen) on doit considérer que cette proposition est indécidable. Mais l'influence de cette hypothèse sur les mathématiques est restée à peu près nulle.*	Since the work of the American Cohen in 1966, this proposition must be regarded as undecidable. But the influence of this hypothesis on mathematics has remained negligible.

E·34 COMMENT That last remark is an echo of the pronouncements of Dieudonné and of Godement on the foundations of mathematics. It is objectionable because they are using "true" relative to some set of axioms, which they might change at will, but pretend that "true" has some absolute meaning.

E·35 COMMENT 1966 is the date of Cohen's book [Coh2]: the news of his discoveries, which only reached me as a Cambridge undergraduate in 1964, first broke in late 1962; two formal announcements followed [Coh1]. Precise dating will be found in [Ka], which accurately portrays the atmosphere of excitement created by Cohen's break-through.

E·36 REMARK On page 38, there is an unsignalled use of AC in the proof of Corollary 2, about the countability of the union of a countable family.

E·37 REMARK The model for their axioms in which unordered pairing fails, presented in [M10], refutes the contention at the end of [Bou49] that the system presented, Bou49, suffices for all the mathematics "of the present day"—even in 1949 one would have wished to prove that for each c and d, the set $\{c, d\}$ exists—and, since $\{c, d\} = \{c\} \cup \{d\}$, confirms the misgivings of Rosser,[12] who suggested in his review [Ro1] that the existence of $a \cup b$ for two arbitrary sets a, b, would not be provable in Bou49.

E·38 HISTORICAL NOTE Entries in *La Tribu* show that Bourbaki consulted Rosser more than once in the early 1950's: it might be that members of Bourbaki contacted him following the appearance of [Ro1], but there may

---

[12] J. Barkley Rosser, 1907–1989; Ph.D. Princeton 1934; at Cornell, 1936–1963; at Madison from 1963.

have been earlier contact, as a paper by Rosser on the eliminability of
ι-terms is invoked, but without exact citation, in [Bou49]. Bourbaki and
Rosser might also be linked through the young Halmos, who was in Chicago
from 1946; he had been von Neumann's assistant at Princeton for a spell,
and, some years later, was instrumental, with Rosser, in bringing about
the famed 1957 Logic Summer Institute at Cornell. Professor Derus writes
that he found in a University of Chicago bookstore in 1991 a copy of [Lu]
bearing Rosser's stamp on the title page, and the inscription: "To Barkley
Rosser, a /professional logician, / from an amateur, / Paul R. Halmos /
July 1958."

E·39 COMMENT Jacqueline Lelong-Ferrand and Jean-Marie Arnaudiès are
both *archicubes*, that is, former pupils of the École Normale Supérieure in
the Rue d'Ulm in Paris.$^{\|}$ Madame Ferrand entered that school in 1936 and
in 1939 was ranked first equal with Roger Apéry in the *agrégation masculine*; Dieudonné was a member of the jury and later wrote that "Only two of
the papers impressed me with their sense of analysis and precocious matu-
rity very rare among candidates for the agrégation. Those two were Roger
Apéry and Jacqueline Ferrand." M. Arnaudiès entered the École Normale
Supérieure in the Rue d'Ulm in 1960. I am told that no logic was taught
there in the early sixties, nor was any logic taught at the Rue D'Ulm when
Madame Ferrand was there in the 1930's, partly because of the tragically
premature death of Jacques Herbrand in a mountaineering accident. The
main subjects then being taught were differential geometry, mathematical physics and probability, the protagonists being Elie Cartan, Louis de
Broglie and Georges Darmois. Where, then, did our authors imbibe their
particular view of logic? Who taught them? It would seem that they both
originally learned the system of Bourbaki's 1949 address, but that at some
point awareness of the shortcomings of that system had filtered through,
leading them to hint that certain unstated axioms are needed to supplement
those stated.

E·40 HISTORICAL NOTE Indeed it appears that there was no teaching of
mathematical logic at the Rue d'Ulm until 1989, since when it has been
maintained, at fourth-year level, by a series of mainly three-year contracts:
Jacques Stern 1989-1995; Jean-Louis Krivine 1995-1998; Alain Louveau

---

$^{\|}$ Originally the École de la Rue d'Ulm admitted only men, and the École
de Sèvres only women. Between the two world wars, the Rue d'Ulm admitted
women and men, and a female student could try the entrance examination for
either. Then there was a period in which the Rue d'Ulm returned to admitting
only men. The two Écoles were then merged and from 1986 have formed a single
École Normale Supérieure.

1998-2001; Elisabeth Bouscaren 2001-2004; Patrick Dehornoy 2004-2007; François Loeser 2007-2010; Martin Hils 2010-2013; Zoé Chatzidakis 2013-2016.

At higher levels, the renaissance began in the 1950s with some courses by Jean Ville and Pierre Samuel, and talks in the seminar of algebra and number theory run by Paul Dubreil. While in Europe for his sabbatical year, 1955/56, from Berkeley, Tarski gave five lectures at the Institut Henri Poincaré at the invitation of J.-L.Destouches, who started a logic seminar there, in connection with the Mathematical Physics that he taught at the Faculté des Sciences while teaching Modern Algebra at the École Centrale; his assistants were Daniel Lacombe, Jean Porte and Roland Fraïssé.[*] The development of French logic was further helped by the two years, 1960/62 that Georg Kreisel spent in Paris, invited by Henri Cartan to lecture at the Sorbonne. Tarski returned in 1962 to lecture at Clermont-Ferrand. Roger Martin taught logic in the philosophical faculty at the Sorbonne from 1964, and at Paris-V from 1969 till his death ten years later.

---

[*] Information from French colleagues, who further write: *Lacombe—le premier logicien à avoir une poste dans une Faculté des Sciences—est le premier spécialiste français de la théorie des fonctions récursives et oriente rapidement des étudiants vers l'emploi des ordinateurs et le langage Lisp pour tester des hypothèses. Porte, formé par Jean Ville et chargé de l'exploitation d'un ordinateur de l'institut Blaise-Pascal, enseigne à la fois la statistique et la théorie des systèmes formels. Fraïssé, formé par l'ancien Bourbachiste René de Possel, enseigne le calcul des formules logiques et la théorie des relations.*

F: *... that has, I suggest, led to the exclusion of logic
from the CAPES examination.*

I N THE FRENCH EDUCATIONAL SYSTEM, there are collèges for pupils
aged 12–15 and lycées for pupils aged 16–18. If you wish to teach in
a collège you must have done successfully a third-year university course,
called a licence, and obtained the *Certificat d'Aptitude au Professorat de
l'Enseignement du Second Degré*, commonly called the CAPES, for which
in 2007, there were 5388 candidates in mathematics, of whom 952 passed;
a success rate of about 18%. To teach in a lycée you must have done successfully a fourth-year university course, called a maîtrise and pass another
examination called the agrégation. The written part of each of these examinations is on a syllabus specified by a national committee; and these
syllabi serve as paradigms for the content of university licence and maîtrise
courses, since universities seeking to attract students for these courses naturally wish to provide teaching on topics for the examinations CAPES and
agrégation, success in which is the aim of perhaps the bulk of those students.

Thus it comes about that the syllabi for CAPES and agrégation have
a profound influence on the whole educational system, and naturally a
uniformising influence.

Come with me now to examine the syllabus for the CAPES, which for
a given year is announced in April or May of the preceding year in a special
number of the *Bulletin Officiel*; in many years details for some subjects
are not given explicitly, but merely stated to be the same as in a previous
year. A complete statement in the *Bulletin Officiel* of the programme for
mathematics was published on 24 May 2001 (for the session of 2002) in
Special Number 8, pages 112–124, to which some minor modifications to
the section on algebra and geometry were published on 20 May 2004 in
Special Number 5, pages 57–58.

F·0 In the *Bulletin* of 2001, the programme is divided into four sections;
three of the sections are further divided into chapters, as follows:

1- Notions sur la logique et les ensembles
   I. Généralités sur le langage et le raisonnement mathématiques. Éléments
   de logique.
   II. Ensembles, relations, applications.
  III. Rudiments de cardinalité.

2- Algèbre et géometrie
   I. Nombres et structures
   II. Polynômes et fractions rationelles
  III. Algèbre linéaire

IV. Espaces euclidiens, espaces hermitiens

V. Géometrie affine et euclidienne

3- Analyse et géometrie differentielle

I. Suites et fonctions

II. Fonctions d'une variable réelle: calcul différentiel et intégral

III. Séries

IV. Équations differentielles

V. Notions sur les fonctions de plusieurs variables réelles

VI. Notions de géometrie différentielle

4- Probabilités et statistiques

Section 4 is not divided into chapters. In all sections there is a further subdivision into paragraphs, which contain lists of topics. In the fourth section, one topic is "Parallèle entre le vocabulaire probabiliste et le vocabulaire ensembliste à propos des opérations sur les événements."

In Sections 2, 3 and 4, the title of the section is immediately followed by the title of the first subdivision; but at that point in Section 1, on page 112, there is inserted the minatory sentence:

Tout exposé de logique formelle est exclu.

That sentence is also to be found in the *Bulletin Officiel* Special Number 3, of 29 April 1999, (the earliest year accessible to me), on page 97; and was left unaltered for ten years by subsequent *Bulletins Officiels*.[37,38]

F·1 Thus there was, officially, a ban on formal logic in each of the sessions 2000-2009, and the ban, though now muted, continues. The objection seems to have been to the actual process of formalisation, for the topics listed in detail in paragraph I of section 1 form an entirely reasonable and coherent group, though their ordering might be challenged, as discussion of the distinction between free and bound variables is placed before discussion of the propositional calculus; and of course the subsection on probability will perforce contain much set theory and Boolean logic. Is it too far-fetched

---

[37] Special Numbers 4, 18 May 2000, page 72; 13, 30 May 2002, page 41; 3, 22 May 2003, page 87; 5, 19 May 2005, page 123; 3, 27 April 2006, page 138; 3, le 17 May 2007, page 122; and 4, 29 May 2008, page 123.

[38] In the annual reports of the Jury of the CAPES from 2003 onwards the minatory sentence is, interestingly, replaced by the more delicate disclaimer that *Aucun exposé de logique formelle n'est envisagé*; which replacement has now been made in the very text of the CAPES mathematical syllabus itself, as witness its recent complete statement in Special Number 6, 25 June 2009, of the *Bulletin Officiel*. But, that possible softening and typographical improvements aside, the 2009 syllabus hardly differs from that of 2001.

to suggest that the origins of this nervousness about formalisation, and by extension, about logic, is the Bourbachiste confusion over quantification?

F·2 I hear (though may have difficulty in verifying, given the secrecy of much French decision-making) that the ban on logic was imposed by a CAPES committee comprised largely of disciples of Bourbaki.$^\diamond$ I am not privy to their secrets, and of the educational background of *Inspecteurs Généraux de Mathématiques* know only that some were themselves *archi-cubes*, and therefore can only guess that the committee's action is rooted in the distrust of logic evinced by Godement, Dieudonné and their colleagues, as documented in Sections C, D and E, and not dispelled by other potential educational influences.

We have seen that in the nineteen-tens and twenties, there was a widespread effort to develop predicate logic. Hilbert made a proposal which was developed by his disciples and cast in concrete by Bourbaki; so that Bourbaki's foundational ideas are rooted in logic as understood by followers of Hilbert in the nineteen-twenties. Hilbert's proposal is not the only possible treatment of predicate logic, but must be one of the clumsiest as measured by the lengths of formulæ generated. We have seen in Section B various idiosyncrasies of the τ operator and its unsuitability in a formalism for set theory, and the inadequacy for functorial ideas of the concomitant treatment of classes.

Thus it would seem that the spiritual ancestor of all the oddities that I dislike in Bourbaki's treatment of logic and set theory is that 1922 paper of Hilbert. The story seems to be that a great man became interested in an alternative but non-optimal approach to a problem; then, rather like newly-hatched ducklings, the early Bourbachistes followed him in adopting that approach; and when, later, they came to positions of influence and power, caused others to do likewise, even though the problem itself had by then been shown to be insoluble.

Bourbaki, in short, by uncritically taking the Hilbert–Bernays–Ackermann operator as their cornerstone, created a nightmare. By working with a formalism that is the product of the enthusiasms of a pre-Gödelian age, they arrived at a negative view of logic and thus created in the minds of their readers a barrier against understanding the aims and enthusiasms of post-Gödelian foundational studies. But what they ought to have been hostile to is not logic but their own (or rather Hilbert's 1922) twisted version of it.

$\diamond$ Is it coïncidence that the other area of mathematics subject to considerable reservations in the CAPES syllabus is probability and statistics, whereas one learns from the embarrassed reminiscence of Laurent Schwartz quoted by Maurice Mashaal [PlS, p.76] that probabilists were subjected to numerous public insults from followers of Bourbaki?

F·3 For despite their fixation on the Hilbert–Bernays–Ackermann operator, Bourbaki did at least notice that logic based on it is seriously flawed, though they might not consciously have identified the nature and cause of the flaw. Rather, they appear to have reasoned that "Hilbert was a great man; his treatment of logic is messy; therefore logic is a mess." Section C shows how a member of Bourbaki betrays in his remarks about logic and set theory numerous apprehensions and misapprehensions, and Sections D and E identify further coarsenings of the situation.

One would expect the degeneration that I have described to have been the subject of comment by concerned mathematicians: but not a peep; it appears that people are frightened to speak out. The Bourbachiste œuvre constitutes a remarkable achievement, and the members of Bourbaki are individually so distinguished, each in his own sphere of competence, that one has to be extremely careful in criticising them; nevertheless I contend that they have infected mathematicians across many generations with their stunted conception and phobia of logic, and that this regrettable result is the consequence of their inadequacy as logicians coupled to their eminence as mathematicians.

F·4 So the CAPES committee may well, in the short term, have made a realistic decision: distinguished and widely-read textbooks of algebra such as the two we have examined present accounts of logic that are repellent; so it is hardly to be expected that in teacher-training colleges, known in France as IUFMs, logic will be well-taught; so one can see why it might have been thought desirable to bar it from the examination.

F·5 But that policy in the long term will, I submit, be intellectually crippling. These syllabi govern to some extent the subjects that can be taught at University level in France; as most of the 3rd and 4th year students are aiming to teach. **I imagine** that there will be universities in France where no logic is taught, it not being thought necessary, not being in the CAPES, just as senior figures in British universities have been heard to say that "we don't need logicians";♡ therefore the next generation of schoolteachers will be giving their pupils a view of mathematics without formal logic in

---

♡ To give this matter some international perspective: in 1984/5, when Cambridge was considering increasing the amount of teaching offered there of logic, including set theory, model theory and recursion theory, statistics from thirteen American universites—Berkely, Boulder, CalTech, Chicago, Cornell, Harvard, Madison, M.I.T., Penn State, Princeton, Stanford, UCLA, and Yale—were produced concerning the number of logic lectures offered to undergraduates and to graduates at these places. Taking as a unit the Cambridge standard lecture length of 50 minutes, it was found that for undergraduate teaching the mean and variance among those thirteen were 93.23 and 32.06 units respectively; at the time

it. Indeed that particular piece of "dumbing down" has already happened: mid-career French mathematicians tell me that as fourteen-year-olds they were fascinated to be introduced to truth tables and formal reasoning, but that today introductory logic is no longer taught in French schools.

That in turn will lead to mathematics itself not being taught in many schools, as, I am told, is already the case in the educational systems of certain countries. So it would be much better to replace all that warped account with a sensible and correct account of logic and set theory.

F·6 REMARK What we have seen is an example of the "trickle-down" process in learning. A bad decision at research level leads in turn to bad teaching at university level, to bad preparation of school teachers, and to bad teaching at school level; thus the scorn for logic displayed by Dieudonné to Quine in a Parisian seminar became, some twenty years later, an entrenched global policy of the French educational system; with the result that French schoolchildren today are described as being *angoissés* by mathematics.

F·7 REMARK Beyond the CAPES is a higher examination called the Agrégation, success in which guarantees a teaching appointment in a *lycée*. Until recently, the syllabus for the Agrégation had not a word about logic, and was arranged under these headings:

I-Algèbre linéaire
II-Groupes et géometrie
III-Anneaux, corps, polynômes et fractions rationelles
IV-Formes bilinéaires et quadratiques sur un espace vectoriel
V-Géometrie affine, projective et euclidienne
VI-Analyse à une variable réelle
VII-Analyse à une variable complexe
VIII-Calcul différentiel
IX-Calcul intégral et probabilités
X-Analyse fonctionelle
XI-Géometrie différentielle

But fortunately the need of computer science departments for logic courses of a particular kind has led to a revision of this syllabus. So the rigid stance portrayed above is, encouragingly, beginning to be modified; but how did it come about in the first place?

---

the offering in Cambridge was 16 units, and, it seems, in Paris 0. For graduates, the American mean and variance were 179.31 and 67.19 units; in Cambridge the amount varied annually, but averaged perhaps 36. What was available in Paris?

G: Centralist rigidity sustains the confusion and consequently flawed teaching; ...

WHEN AN EDUCATIONAL SYSTEM is prescriptive, what is to be taught will be laid down at the centre, thus negating the personal and individual nature of teaching. The undesirability of that is excellently expressed in the following passage from Feyerabend, *Against Method* [Fe]:

p 45: "*Any method that encourages uniformity is in the last resort a method of deception. It enforces an unenlightened conformism and speaks of truth; it leads to a deterioration of intellectual capabilities, and speaks of deep insight; it destroys the most precious gift of the young — their tremendous power of imagination — and speaks of education. Variety of opinion is necessary for objective knowledge.*"

The legacy of Napoleon: the foundation of the modern French university system

G·0 French *dirigisme*—the taste for strong orders from the centre—goes back at least to Richelieu; in 1789 as the monarchy tottered a prescient nobleman remarked that "If the king will not have an army, the army will have a king," and within ten years a young general from Corsica had risen to supreme political power.

G·1 Since coming to work in France I have found that in order to understand the way French academics behave, one should imagine that one is in the army; and it has seemed to me that there are no universities in France, only units in a university system. The reason emerged in 2008: the University from which is descended the contemporary French educational system was founded by decree two centuries ago, when the Emperor Napoléon I caused the Corps Législatif to pass the decree/law of May 10 1806, visible at

http://www.inrp.fr/she/universite_imperiale_bicentenaire_loi.htm

which enacted that:

Art.1er. *Il sera formé, sous le nom d'Université impériale, un corps chargé exclusivement de l'enseignement et de l'éducation publique dans tout l'Empire.*

Art 3. *L'organisation du corps enseignant sera présentée, en forme de loi, au Corps législatif, à la session de 1810.*

But the Emperor could not wait so long and hurried things forward by a long and detailed decree of March 17, 1808, visible at

http://www.inrp.fr/she/universite_imperiale_bicentenaire_decret.htm

which enacted that:

Art 1er. *L'enseignement public, dans tout l'empire, est confié exclusivement à l'Université.*

Art 2. *Aucune école, aucun établissement quelconque d'instruction ne peut être formé hors de l'Université impériale, et sans l'autorisation de son chef.*

Art 3. *Nul ne peut ouvrir d'école, ni enseigner publiquement, sans être membre de l'Université impériale, et gradué d'une de ses facultés. Néanmoins, l'instruction dans les séminaires dépend des archevêques et des évêques, chacun dans son diocèse.*

Art 4. *L'université impériale sera composée d'autant d'académies qu'il y a de cours d'appel.*

This document makes fascinating reading: of the 144 articles, I draw attention to

Article 5, which shows the comprehensive character of the concept: all educational establishments, down to "Dames' Schools", are to come under the single umbrella;

Article 29, which fixes the rank of the various fonctionnaires: the university professors are at level 10; above them are the Grand Master, the Chancellor, the Treasurer, and assorted councillors, inspectors, rectors and deans;

Article 38, which states that the base of the teaching at all levels is to be fourfold: the precepts of the Catholic religion; fidelity to the emperor and his dynasty; obedience to the statutes, of which the aim is to create citizens that are attached to their religion, their prince, their country and their family; conformity to the dispositions of the Edict of 1682 concerning the four propositions contained in the declaration of the clergy of that year;□

Article 101, which proposes something like an Oxbridge college in stating that teachers of junior rank will be constrained to be celibate and to live in community; more senior professors may be married, but if they are single they are encouraged to "live in" and benefit from communal life;

Article 102, which provided that no woman is to be lodged or received in any lycée or collège; and

Articles 33, 128, 129, 130, which concern the robes to be worn by members of the various Faculties.

A professor is required, by clause 8 of Article 31, to be a doctor of his Faculty. To become a doctor in, for example, the Faculty of mathematical

---

□ the purpose of which was to limit the authority of the Pope in France and in effect to create a Church of France somewhat similar to the Church of England.

sciences and physics, one must, by Article 24, submit two theses, in those subjects which one intends to teach. Mention of such theses aside, the Imperial decree, though frequently speaking of *le corps enseignant*, the teaching body, says not a word about research; but what might Napoleon, a soldier of genius but not an academic, be expected to know about that?

Politics and mathematics

G·2 Despite the ceaseless tug-of-war between the pro- and anti-clerical parties over the educational system in France in the 19th century and the various perturbations in the 20th, the centralist conception of Napoleon remains. In our present discussion we see French *dirigisme* at work in the CAPES. I wonder how similar is this imposition of a regressive policy by an uncomprehending bureaucracy to the situation of logic in Eastern Europe under Stalin, described in my essay *Logic and Terror* [M2], when, for example, in Poland the great Andrzej Mostowski dared not call himself a logician till the late nineteen-sixties.

G·3 When in 1871 the Third Republic began and Jules Ferry took control of the educational system, his concern was to ensure that only republican ideas would be taught. Try as I might, I can see no relationship between republican values and mathematics, but the French can; and, a hundred and forty years later, a section on "republican values" was included in the syllabus recently proposed in Réunion for the second year Master course in mathematics, which syllabus also mentions developments "since 1789". Is this an echo of the course on *arithmétique républicaine* taught in Rouen in 1794 by Caius-Gracchus Prud'homme?

G·4 A reader of *The Ignorance of Bourbaki*, the holder of a (C4) chair of pure mathematics at a leading German University, told me that as a young man he had been reduced to a state of intellectual paralysis by reading Bourbaki and that he had had to retire from mathematics for six months before making a fresh start. Fortunately, it is not necessary to worship at the Bourbachiste shrine in order to do serious mathematics.

G·5 That it might ever have been thought so necessary can be divined from fleeting remarks about intellectual terrorism by Miles Reid in his book [Re2]. I quote from the remarks on pages 114–117 of the 1994 reprint.

> *"Rigorous foundations for algebraic geometry were laid in the 1920s and 1930s by van der Waerden, Zariski and Weil. (van der Waerden's contribution is often suppressed, apparently because a number of mathematicians of the immediate post-war period, including some of the leading algebraic geometers, considered him a Nazi collaborator.)"*

"By around 1950, Weil's system of foundations was accepted as the norm, to the extent that traditional geometers (such as Hodge and Pedoe) felt compelled to base their books on it, much to the detriment, I believe, of their readability."

"From around 1955 to 1970, algebraic geometry was dominated by Paris mathematicians, first Serre then more especially Grothendieck."

"On the other hand, the Grothendieck personality cult had serious side effects: many people who had devoted a large part of their lives to mastering Weil foundations suffered rejection and humiliation. ... The study of category theory for its own sake (surely one of the most sterile of all intellectual pursuits) also dates from this time."

"I understand that some of the mathematicians now involved in administering French research money are individuals who suffered during this period of intellectual terrorism, and that applications for CNRS research projects are in consequence regularly dressed up to minimise their connection with algebraic geometry."

G·6 Let us set against Reid's remarks a comment of Armand Borel in [Bor]:

"Of course there were some grumblings against Bourbaki's influence. We had witnessed progress in, and a unification of, a big chunk of mathematics, chiefly through rather sophisticated (at the time) essentially algebraic methods. The most successful lecturers in Paris were Cartan and Serre, who had a considerable following. The mathematical climate was not favourable to mathematicians with a different temperament, a different approach. This was indeed unfortunate, but could hardly be held against Bourbaki members, who did not force anyone to carry on research in their way."

G·7 COMMENT I wonder if there is an element of complacency in that last statement of Borel. Suppose it were the case that over a certain period in numerous universities, in France, in Spain, in England, or elsewhere, the Bourbachistes seized power and pursued a policy of denying jobs to non-Bourbachistes. How would one obtain evidence of that? The poor non-Bourbachistes, being excluded from employment which would permit them to research would be likely to move away from universities and find jobs in industry or elsewhere, and indeed to lose touch with research mathematics. So they would be excluded from any figures that might be produced. People would be saying that the Bourbachiste view is the standard one; what would not be said is the subtext, that that state of affairs has come about because the opposition has been suppressed. There would thus be a political component to what has been called mathematical practice.

An explicit example: François Apéry writes in [Ap1,2] of his father Roger Apéry, who at the age of 61 proved the irrationality of $\zeta(3)$, that he declined Dieudonné's invitation to join Bourbaki, and that the dominance of Bourbaki meant marginalisation for an anti-Bourbakiste; despite, one might add, Apéry *père* having at the age of 23 impressed Dieudonné by his performance in the *agrégation masculine*.[39]

G·8 That prudent would-be critics of Bourbaki should conceal their identity is suggested by the circumstance that in the special issue [PlS] of *Pour la Science*, published in 2000, dedicated to a study of Bourbaki, on page 78, where I am named and briefly quoted, and described, to my delight, as having *pourfendu l'ignorance bourbachique*, a *spécialiste parisien* is quoted at greater length but **on condition of anonymity.**

G·9 To that Parisian critic's remarks, with which I am in complete agreement, I would add that it is not only in France that Bourbaki is regarded as an unchallengeable authority on logic. Their stifling influence is to be found elsewhere. Let me give a comparatively mild example. A mathematical logician has confided in me that he obtained tenure at his University, in a European country other than France, by pretending that despite retaining an eccentric interest in logic, in reality he subscribed to his Department's view that "real men don't do logic". He believes, and I with him, that had he revealed the depth of his commitment to logic he would not have been given tenure, for the reason that the quasi-totality of his decision-making colleagues were imbued with Bourbaki's negative attitude. I could wish that now that he has landed safely in the Realm of the Blessed, he would speak up for logic, but it appears that the habit of caution is too deeply ingrained. Still, it is not for me to "out" him.

G·10 Professor Segal in his *Zentralblatt* review [Seg1] of my essay [M3] writes that I am unhappy with the neglect of logic by mathematicians. No, it is not the neglect — surely all are free to be as ignorant as they choose — to which I object but the imposition, by the high-placed ignorant, of their ignorance on their subordinates, their interference with the teaching of logic to those who wish to learn it, and their denial, through the mechanism mendaciously called "peer review"$^\heartsuit$, of research funds for work in this area.

G·11 COMMENT In 1851, at the Albany meeting of the American Association for the Advancement of Science, Alexander Dallas Bache, in his address as out-going president, spoke of *that modified charlatanism which*

---

[39] See [Ag] for further references and discussion, and [Ch] for a public protest by Chevalley and others against the neo-Bourbachiste cosily repressive attitude lampooned by Molière: *Nul n'aura de l'esprit hors nous et nos amis.*

$^\heartsuit$ "Clique review" would be more accurate.

makes merit in one subject an excuse for asking authority in others, or in all. In 1974, in his Nobel Memorial Prize address *The pretence of knowledge,*[40] Friedrich August von Hayek describes the "scientistic" attitude as *decidedly unscientific in the true sense of the word, since it involves a mechanical and uncritical application of habits of thought to fields different from those in which they have been formed.*

Somewhere between those two is what, in a nutshell, I fear has happened with Bourbaki and logic.[41]

---

[40] In English in [vHay1], in German translation in [vHay2].

[41] The point is reïnforced by the intriguing lecture [Du] given by Till Düppe at Siena in October 2007, entitled *Gerard Debreu from Nicolas Bourbaki to Adam Smith,* exploring the harmful consequences for economics of Bourbaki's influence on the psychological relationship of mathematical economists to their subject.

H: *... the recovery will start when mathematicians adopt a post-Gödelian treatment of logic.*

THE MANY DEVELOPMENTS IN LOGIC and foundational studies since the 1920's—it would be invidious to name names, but let me mention the incompleteness theorems, the relative consistency proofs for the Axiom of Choice, the evolution of category theory, in set theory the discovery and development of forcing, the emergence of large cardinal properties, the fine structure and core model programmes, and the work on infinitary games, to say nothing of the many advances in model theory, proof theory and recursion theory—have created a foundational arena so different from that envisaged in the 1920's as to demand a new and positive approach to the teaching of logic in schools and universities.

Concerning that teaching, I would say that *of course* there are things in logic which are not yet understood, as in any living subject that tackles difficult problems, but there are ways of presenting logic and set theory which are far better than the Bourbachiste method. My message to the adventurous youth of today must be this: if you want to know what has happened in logic in the past century, do not go to Bourbaki, for they cannot tell you.

H·0 There is a widely held view, vehemently urged by Dieudonné and spread by the less enlightened of Bourbaki's disciples, that mathematicians need trouble themselves no longer about foundational questions. But there are many examples of classical conjectures being proved both consistent and independent: one might mention Souslin's hypothesis about a possible characterization of the real line as an ordered set, and Whitehead's conjecture concerning free Abelian groups; one might also mention the use of set theory in elucidating the structure of weakly distributive Boolean algebras and in the study of the Lebesgue measurability of an arbitrary set of reals. And if one looks only for positive contributions of ideas from logic to other branches of mathematics, one finds that they too are legion. Immediately to mind, again without naming names, come the use of model theory in the proof of the near-truth of Artin's conjecture and in the proof of the Lang–Mordell conjecture for fields of arbitrary characteristic; the use of Ramsey theory in the study of the subspace structure of Banach spaces and in the positive solution of Kuroš' problem concerning the existence of uncountable groups with only countable subgroups; the use of priority arguments from recursion theory in the construction of topological manifolds; and the use of proof theory in the study of sums of squares.

So I think mathematicians would be unwise to tell themselves that they will never encounter foundational problems nor have a use for foundational ideas.

Bourbaki and French nationalism

H·1 In my essay *The Ignorance of Bourbaki* I wondered whether the attitudes of Bourbaki might stem from the influence of Hilbert or from some nationalist or chauvinist feeling, and Professor Segal, in his review [Seg1], suggested that I was thereby contradicting myself.

Perhaps I should state that I see a distinction between nationalism and chauvinism. Consider, for example, Janiszewski, who at the end of the First World War called for a small poor country to make its mark in foundational studies: I see him as a Polish nationalist but not a chauvinist. It is one thing to say "Good things are going on elsewhere in the world: let us try to do as well or better." It is another to say "Everything that is worth knowing is known by us; let us ignore the activities of others".

H·2 Now Cartier's interview [Sen] makes it clear that Hilbert and German philosophy were held up as models by Weil and others. He says

> "The general philosophy is as developed by Kant. Bourbaki is the brainchild of German philosophy. Bourbaki was founded to develop and propagate German philosophical views in science. All these people ... were proponents of German philosophy."

H·3 So I really do not see that there is a contradiction between wishing to strengthen French mathematics and saying that the Germans do it better. One might say that the Bourbachistes were nationalist but not chauvinist. They considered, indeed, that the French policy of putting scientists in trenches in World War I, when the Germans, wisely, protected their scientists, had retarded French mathematics by one full generation. Further evidence comes from *Claude Chevalley described by his daughter*, [Cho, pages 36–39], where she says that the Bourbaki movement was started essentially because rigour was lacking among French mathematicians by comparison with the Germans, that is, the Hilbertians.

The chimera of completeness

H·4 Thus we come back to Hilbert. We began this essay with a translated excerpt of a letter from him to Frege. I suggest that Hilbert never shook off the illusion that a complete recursive axiomatisation of the whole of mathematics awaited discovery. It underlies his championing of the epsilon operator, the use of which seems to rest on a belief in the completeness of the system under discussion.

It reappeared in his quarrel [vDa1] with Brouwer: from the sources quoted in [Ke] it is evident that Brouwer had accepted some perhaps intuitive notion of the incompleteness of mathematics, and indeed Brouwer's lectures in Vienna may have influenced the young Gödel to probe further. Such a perception was anathema to Hilbert, whose battle-cry was *Wir müssen wissen, wir werden wissen.*

H·5 Bourbaki also were bedevilled by this mistake. Though Bourbaki in their *Note Historique* give a correct sketch of a proof of the incompleteness theorems, their residual identification of truth with provability produces problems. In a complete system, truth and provability are indeed identical; but they are not for recursively axiomatisable consistent systems extending arithmetic. Bourbaki, indeed, are incoherent in their use of "true". They have the idea that it is relative to a certain system; but they also at times wish to say that "true" means "known to be provable at the time of writing".

H·6 A long-held belief, by the time it is shown to be false, may be too deeply embedded to be given up. We saw Hilbert, in his preface to [HiB1], doggedly maintaining that his programme would survive Gödel's unwelcome discovery of incompleteness. Corry, in [Cor3] and [Cor5], shows *inter alia* that Hilbert discarded the view that mathematics is a formal game with marks on paper. Hilbert himself developed his ideas about logic over twenty years or more, as the subject itself developed through the work of many people, leaving us with the problem of explaining Bourbaki's strangely rigid, and indeed oppressive, attitude to logic.

H·7 Part of that oppressiveness may stem from the fact that Bourbaki was not a person but a group of people, so that the mind of Bourbaki is not an entity of the same kind as the mind of Hilbert, and would be subject to discontinuities in its development stemming from tensions between individual members of the group. The archives of Bourbaki are, at least in part, available on-line,[42] at

http://mathdoc.emath.fr/archives-bourbaki/feuilleter.php

They make poignant reading. The numerous drafts, by different hands, of the various sections of the book on set theory make manifest the extent of the effort that went into the preparation of the finished work.[43] Among those drafts, my eye is caught by two typescript notes by Chevalley entitled *Ensembles bien ordonnés* and *Le formalisme de Gödel*. These two show that Chevalley was more in tune with mainstream set theory than are

---

[42] I am greatly indebted to Professor William Messing for this information.

[43] The names pencilled on copies of drafts prior to their distribution are of recipients rather than authors.

Bourbaki's books, and that he had gone so far as to read[44] at least part of Gödel's monograph on the axiom of choice and the generalised continuum hypothesis.* Had Chevalley's voice been louder than Dieudonné's in the shouting matches that, Armand Borel tells us, were a regular feature of Bourbaki meetings, perhaps more of the insights of Gödel would have got through the Bourbaki process, and the volume on logic and set theory would have been far more satisfactory, with fewer of its readers coming to feel that they had, in some sense, been cheated. When Bourbaki decided against the approach to set theory that Chevalley, following his reading of Gödel, would, apparently, have favoured, the result was to create a breach between mathematics as conceived by Bourbaki and set theory as developed by Gödel, and, following Cohen's breakthrough, by Solovay, Jensen and their successors. If only ...

La Tribu

H·8 But even more revealing than the drafts are the issues of *La Tribu*, the Bourbaki in-house journal that contained minutes of the various meetings, including commitments for the future, censures given to various members who had come to meetings ill-prepared, and, here and there, some excellent parodies and jokes. Hitherto in this essay we have treated, as we must, the works of Bourbaki as they were actually published, not as they might have been; but the copies of *La Tribu* enable us to penetrate beneath the surface and find out something of the personal interactions that led to the final choice of treatment.

One cannot know how faithfully the minutes of *La Tribu* recorded the discussions of the meetings; nevertheless it is noteworthy how many points made in earlier sections of this essay against the published logic texts of Bourbaki and their followers are reported to have been the subject of debate at their meetings. We cannot here examine *all* pertinent passages, illuminating though it be to observe the developing perception among members of the group of a need for a book on logic and set theory; to note the adoption in 1950 of the Hilbert operator; and to note the 1951 decision, apparently Dixmier's, to replace the Lesniewski–Tarski system of propositional logic by that of Hilbert–Ackermann, *"arrangé à la sauce Chevalley"*; we shall

---

[44] As indeed did the young Cartier, who thereby was enabled to have a serious and lengthy conversation with Gödel himself, in German, at Princeton in 1957.

* There are, though, inaccuracies in Chevalley's summary of Gödel's consistency proof: he fails to distinguish between the class of all subsets of a constructible set and the class of all its constructible subsets, and thus appears in places to think that every subset of a constructible set will itself necessarily be constructible.

focus particularly on two further people, Eilenberg and Rosser, their views of logic and their relations with Weil.[45]

H·9 The reader of *La Tribu* will note the interest, expressed repeatedly, in the foundational proposals of Eilenberg, who is usually called Sammy. Eilenberg (1913–1998) had taken a logic course from Tarski in Warsaw in the early 1930's; at his father's urging, he left Poland for the U. S. in 1939, where he was received at Princeton and then appointed to Michigan.

He was recruited into the Bourbaki group at the instigation and urging of Weil as a result of Weil's esteem for his work as a topologist.[46]

From later conversations with Eilenberg, Cartier had the strong impression that although Eilenberg indeed knew more about and took more interest in logic than other Bourbaki members, he was never very interested in its role in foundations or even as mathematical hygiene but viewed it more as one further subject matter to which current abstract structural mathematics could bring deeper methods and concepts.[47]

Eilenberg in conversation with Michael Wright in 1990 said that he thought of foundations as something growing and evolving along with the main body of mathematics, concerned mainly with tracing the relationships within that and as *"something coming into focus as we move from the inside outwards as mathematics grows and we come to see how the various directions of advance are connected"*.

TRIBU 15: Compte-rendu du congrès de Nancy (9 au 13 avril 1948)
[nbt_017.pdf][48]
PRÉSENTS: Chabauty Delsarte Dieudonné Godement Roger Samuel Schwartz Weil

Mornings at this meeting were devoted to Livre I. Its discussion opens with a revelatory pleasantry:

---

[45] My discussion of the parts played by Chevalley, Rosser, Eilenberg and Dixmier owes much to the illuminating reminiscences of Pierre Cartier in conversation with Michael Wright in Paris on January 9th 2012.

[46] Cartier gives another example of Weil's influence: he says that the collaboration, which in 1949–50 was already under way, of Eilenberg and Cartan in what became their famous text, was very much imposed—at the beginning—by Weil in the face of initial reluctance from Cartan. Cartan at that point would have preferred to write his own text single handed, although in the course of the collaboration he developed a great appreciation for all that Eilenberg brought to the work.

[47] This attitude clearly marked Eilenberg's student Lawvere.

[48] The numbering of the on-line .pdf files is slightly out compared with the numbering of the issues of *La Tribu*.

p 5 *Malgré le soin constant de chacun de ne pas faire de philosophie ("Phi-losophy is the systematic misuse of a language especially created for this purpose"), le Congrès fut souvent menacé d'enlisement. Dieudonné rappela à l'ordre les récalcitrants, et "ontologiste" fut l'injure suprême. On nota une éclatante conversion de Schwartz à la Dialectique.*

There follows a lengthy discussion of the difficulties to be encountered at the start of an exposition of logic. Weil is instructed to re-write the Introduction taking account of the discussion.

TRIBU 18: Congrès oecuménique du cocotier (Royaumont, 13 au 25 avril 1949)   [nbt_020.pdf]
PRÉSENTS: Cartan Chevalley Delsarte Dieudonné Godement Pisot Roger Samuel Schwartz Weil, et le COBAYE Serre (en cour de métamorphose).
ABSENTS: Chabauty, Ehresmann

p 1 *Soucieux de l'avenir, Bourbaki décida d'envisager des situations de repli pour ses membres; la liste suivante a été adopté:* and then on page 2, Eilenberg is in the list, proposed as a *concierge* in a *collège de filles,* although his first attendance at a meeting seems to have been at Roy-aumont in October 1950.

p 5 Eilenberg engages to make a report on *ses vieux trucs de multicohérence et d'applications dans les cercles.*

p 6 *Dès la première séance de discussion, Chevalley soulève des objections relatives à la notion de texte formalisé; celles-ci ménacent d'empêcher toute publication. Après une nuit de remords,** Chevalley revient à des opinions plus conciliantes, et on lui accorde qu'il y a là une sérieuse difficulté qu'on le charge de masquer le moins hypocritiquement possible dans l'introduction générale. ...*

p 7 the system of Gödel is mentioned.

TRIBU 19: Congrès de la Réforme (Paris 2 au 8 octobre 1949) [nbt_021.pdf]
PRÉSENTS: Cartan Dieudonné Ehresmann Godement Roger Samuel Serre Schwartz Weil.
COBAYES: Blanchard Malgrange

p 1 Weil proposes that once the contents of a chapter have stabilised, its details should be discussed in committee rather than in plenary session. *La foule applaudit ce projet.* Cartan and Dieudonné ask to be on all the committees. On page 2 the committee for Chapters I and II is constituted as Dieudonné, Cartan and Weil, and for Chapter III as Dieudonné, Cartan and Samuel.

** Was that sleepless night the cause of Chevalley's attending no further meet-ings till July 1952?

TRIBU 20: [Congrès des Comités de Décembre, Paris] 15 Décembre 1949
Compte-rendu des comités de decembre (Paris, 3-5 Déc 1949)
[nbt_022.pdf]

p 1 Weil raises objections to the axiom of families of sets (essentially the axiom of replacement); his chief objection is that Bourbaki would have no use for it;[♣] but the congress repeats *à l'unanimité* its previous decision to reject any axiom system that forbids unrestricted cardinal arithmetic.

TRIBU 22: Congrès de la revanche du Cocotier (5 au 17 avril 1950) Royaumont [nbt_024.pdf]

PRÉSENTS: Cartan Chabautey Delsarte Dieudonné Godement Mackey (au début) Pisot Roger Samuel Schwartz Serre Weil.

p 2 *Pour satisfaire les désirs inavoués de Chevalley, on basera la logique sur le symbole "yoga" de Hilbert.* Is this the first time that the Hilbert ε symbol was discussed? Rosser's review [Ro1] was in the issue of the *Journal of Symbolic Logic* dated January 1950. But it is conceivable that he sent it earlier to Bourbaki and opened a dialogue, and perhaps suggested that they use the epsilon symbol. *Et, on eut beau "chasser le Dénombrable", "il revint en trottant".*

p 4 Drop the idea of publishing chapter III before Chaps I and II.

p 5 Chevalley engages to do chapter II, in particular the section on structures.

p 7 Review of state of Livre I: delete the confessions of Chevalley, and add a justification of the ε of Hilbert.

TRIBU 23:[49] Congrès de l'horizon (Royaumont, 8-15 octobre 1950)
[awt_002.pdf]

PRÉSENTS: Cartan Dieudonné Dixmier Eilenberg Samuel Serre.
RETARDATAIRES: Godement Schwartz Koszul

p 1 *La présence d'Eilenberg fut le fait marquant du Congrès. .... il sera appelé "Sammy".*

La lecture de la logique souleva une indifférence croissante, qui, après Godement, Schwartz et Serre, commence à gagner Cartan et Dieudonné; il fallut l'expulsion de la relation "x est un ensemble" pour faire quelque peu crier Samuel. Seuls Dixmier et Sammy montrèrent un vif intérêt pour ces questions...

p 2 *Engagements du Congrès:*

[♣] This slightly suggests that Rosser's critique [Ro1] of [Bou49] had not yet reached Bourbaki.

[49] Absent from the Delsarte collection, but in the Weil collection.

> DIXMIER: *redige l'état 6 de la Logique et des premiers § des Ensembles.*
>
> SAMMY: *explique à Dixmier son système pour les parenthèses.*

Pages 11-14 contain minutes of an extensive discussion of a version of Chapter I, État 5 that does not always agree in its numbering of sections with that available on-line.

p 11 *On adopte temporairement le symbole ε de Hilbert. ...*

p 14 *Pour le chap.II on rejette la proposition Chevalley de remonter les ordinaux avant les structures.*

> *Cartan voudrait le couple comme relation primitive; d'autres préférer l'astuce Gödel (qui donne aussi l'ensemble à deux éléments). ... Montrer que $\{x, y\} = \{y, x\}$.*

H·10 On 19 January 1951, Weil wrote to Cartan[50]:

"Ci-joint une lettre de Barkley Rosser, commentant mes suggestions sur la logique de Bourbaki. Je te demanderai de la transmettre à Nancy, pour la faire tirer, après en avoir pris connaissance. Chevalley m'écrit qu'il se rallie "avec enthousiasme" (sic!!!) à ma proposition d'employer ε pour définir les entiers et les cardinaux. ....

Je t'envoie deux exemplaires de la lettre de Rosser, un pour Nancy et un pour que tu le transmettes dès maintenant à Dixmier puisque (sauf erreur) c'est celui-ci qui est chargé de la logique."

TRIBU 24:[51] Congrès de Nancy 27 janvier au 3 février 1951 [nbt_025.pdf] PRÉSENTS: Cartan Delsarte Dieudonné Dixmier Godement Koszul Sammy Samuel Serre Schwartz.

COBAYES: Glaesser, Grothendieck, un brésilien

Dixmier engages to finish Draft 6 of logic, for 1 May 1951. On page 3: Dixmier sees contradictions in Weil's suggestions about ε; it is noted that **ceci cadre mal avec les assurances de Rosser.**♠

H·11 Who, one wonders, first suggested using the epsilon operator? Dixmier voiced some reservations; and Cartier recalls Eilenberg saying in the 1960s that though he could see the defects of the Hilbert operator in risking ambiguity of type, that issue did not impinge on the constructions of algebraic

---

[50] [CW], page 327: tantalisingly, the letter from Rosser sent in duplicate by Weil is nowhere to be found, despite the best efforts of, in Strasbourg, the editor, Michèle Audin, of [CW]; in Paris, Florence Greffe, Conservateur des Archives de l'Académie des Sciences; and in Austin, Carol Mead, Archivist of the Archives of American Mathematics.

[51] Erroneously numbered 23 in the Delsarte collection.

♠ Conveyed presumably in the missing letter from Rosser.

topology, so did not justify a fuss. Perhaps Eilenberg felt that since the other Bourbachistes were little interested in logic, and since his relations with Weil were excellent, and helpful to him in his work with Cartan in Topology, which he saw as his main business, it would have done him very little good—especially with Weil—to try to get the others interested.

TRIBU 25: Congrès oecuménique de Pelvoux-le-Poët (25 juin au 8 juillet 1951)   [nbt_026.pdf]

PRÉSENTS: Cartan Delsarte Dieudonné Dixmier Godement Sammy Samuel Schwartz Serre Weil.

ABSENT: Koszul   VISITEURS: Hochschild Borel   COBAYES: Cartier Mirkic

p 3 The plan of Livre I is confirmed as Introduction; I: description of formal mathematics; II: theory of sets; III: ordered sets, integers; IV: structures.

The commitments made by members are listed on pages 4 and 5: in particular Dixmier engages to make the final version of the logic section two months after Rosser gives his imprimatur.

p 5 Sammy: Rapport sur le rôle des foncteurs au Livre I, chapitre des Structures (janvier 52) Samuel promises to write the Introduction to Livre I, (with Weil) by December 1951; and Weil promises to write, with Rosser, the *Note Historique* for Livre I.

On pages 6-9 there is a detailed report on the set theory book; and the decision is there recorded to "send the list of all our axioms to Rosser: If he finds them "kosher", we will proceed immediately to near-final versions." *WEIL a décanulé le contre exemple de DIXMIER sur l'égalité des ε de deux relations équivalentes (voir "Lamentations")*

On page 7, the ordered pair will be taken as primitive. On page 8, it is decided to speak of schemas rather than implicit axioms. On page 9, it is noted that the opinion of Rosser is awaited, and that Weil will seek instruction from him.||

TRIBU 26: Congrès Croupion (1 au 9 octobre 1951)   [nbt_027.pdf]

PRÉSENTS: Cartan Dieudonné Dixmier Godement Samuel Schwartz Serre.

p 3 It is resolved to include the last sentence of the *JSL* article, that—in effect—all can be done in ZC.

p 4 *Rosser trouve kosher notre systême d'axiomes avec l'égalité des τ (remplacera ε pour raisons typographiques) de deux relations équivalents (sous la forme WEIL).*

|| In the original, *"WEIL se fera tapiriser par ROSSER"*. In Normalien argot, a *Tapir* is a schoolchild whose parents pay a Normalien to give him evening lessons to catch up in Maths or in another discipline. A Cambridge translation would be that Weil would ask Rosser for a supervision on logic.

Cartan wants to change Chapter II; so the meeting, feeling baffled, resolves to pursue discussion by letter and in a congress.

TRIBU 27: Congrès Croupion des Vosges, (8 au 16 mars 1952). [nbt_028.pdf]
PRÉSENTS: Cartan Dieudonné Dixmier Godement Samuel Schwartz.
INVITÉ: Grothendieck.

The minutes suggest that, much to the relief of certain others present, Grothendieck returned to Nancy in a huff after being told that all empty sets are equal but some are more equal than others.

TRIBU 28: Congrès de la motorisation de l'âne qui trotte (Pelvoux-le-Poët
         25-6 au 8-7 1952)   [nbt_029.pdf]
PRÉSENTS: Cartan Chevalley Delsarte Dieudonné Dixmier Godement Sammy Samuel Schwartz Serre Weil.
NOBLES VISITEURS ÉTRANGERS: Borel, de Rham, Hochschild

p 2 Livre I: Introduction finished; Chap I adopted; Chapter II: new version of Dieudonné to be examined in October; ditto Chapter III. Chapter IV (Structures): *une nouvelle rédaction sera polie cet automne par un Caucus Americain, puis envoyé au Congrès de Février. Note Historique: Samuel se fera tapiriser par Rosser à Ithaca. On Logic, on a décidé de rédiger l'appendice en style "intuitionistisable" (à la Dixmier). On structures, on décide d'essayer le système Sammy.*

p 5 Eilenberg engages to draft Chapters 1 and 4 of *Topologie Pédérastique*, and Serre to draft chapters 2 and 3 of the same. Samuel engages to prepare Chapter 4 (structures) and the Note Historique.

On pages 7–9 there is a further report on the book *Théorie des ensembles*. The problems addressed by category theory are starting to reveal themselves.

TRIBU 29: Congrès de l'incarnation de l'âne qui trotte (Celles-sur-plaine,
         19-26.10.1952)   [nbt_030.pdf]
PRÉSENTS: Cartan Koszul Serre Weil.   QUASI-PRÉSENTS: Dixmier Schwartz
QUASI-ABSENT: Delsarte

p 4 *Samuel réclamera Rosser pour qu'il donne rapidement son avis sur les chap.I-II.*

Bourbaki consult Rosser

H·12 Thus, behind the scenes, Bourbaki seem to have hungered for reassurance about their foundational book. Rosser appears to have left no record of his meetings or correspondence with Weil or Samuel; nor is there any in the

Weil archives in Paris.[52] Perhaps the sequence of events began in Chicago, whither both Mac Lane and Weil were lured by Stone in 1947; perhaps it was Mac Lane, a founder member, who proposed that Weil be invited to present Bourbaki's foundational ideas to the Association for Symbolic Logic in December 1948; perhaps Rosser's review [Ro1] of [Bou49] led Bourbaki to contact him; perhaps he saw his role as that of protecting Bourbaki from error rather than steering them towards any particular account of logic or set theory. He himself was willing to consider widely differing accounts: he published shortly afterwards a book [Ro2] expounding the development of mathematics within Quine's system NF, of which the review [Cu] by Curry is illuminating; some years later he published a book [Ro3] expounding the Boolean-valued models approach to forcing, in something like a ZF context; and, once,[53] in private and perhaps with provocative intent, declared himself a finitist who disbelieved in the existence of infinite sets.

H·13 Rosser's son has written in [RoJr] of his father's character and achievements. The picture of the notoriously abrasive Weil presenting himself at Rosser's door as a humble seeker after truth refuses to come into focus; but Cartier's view is that Rosser was regarded by Bourbaki as arbiter of last resort in logic solely and simply because that is what Weil proclaimed him to be—Cartier recalls Weil describing Rosser as a good personal friend "who happens to know about logic"—and those members, such as Chevalley, Dixmier and Eilenberg, who knew enough of the work of Gödel and Tarski to doubt this assurance, did not feel there was enough at stake to make an issue of it.

H·14 The friendship between Weil and Rosser rested on some non-mathematical tie or common interest—Weil had a very wide range of interests—which brought them together, perhaps from 1947 when Weil came to Chicago, perhaps from the period before Weil's departure from the U.S. for Brazil in 1945. To quote Michael Wright, given Weil's notorious disesteem for logic, which he was scarcely reluctant to voice, it cannot have been founded on admiration for Rosser's professional achievement as a logician. But Weil always spoke warmly of him.

H·15 The three significant changes to Bou49 that yielded the system finally adopted, Bou54, were:

(H·15·0) to follow the Hilbert–Ackermann treatment of propositional logic;

---

[52] Cartier suggests that the absence of correspondence in the Weil archive may be due to the fact that most of the exchanges would have taken place through Rosser's coming to Chicago to seek out Weil on his home ground.

[53] According to the testimony of Gerald Sacks.

(H·15·1) to treat quantifiers not as primitive signs but as derived from the Hilbert operator;

(H·15·2) to change the set-theoretic axioms from something like ZC to something like ZFC.

I can imagine Rosser suggesting (H·15·0) to Bourbaki, but not (H·15·1), unless in mischief; might the latter have been suggested by Mac Lane, the ex-pupil of Bernays, to Weil, the translator of [Hi2]? If not, perhaps the suggestion came from Chevalley in early 1950, as suggested by Tribu 22, and then was explored by Chevalley in his Draft 5; then adopted provisionally in October 1950 and definitively in Dixmier's Draft 6 in 1951. Let us hope that Rosser's missing letter will re-appear and settle these questions.

H·16 As for (H·15·2), Weil and Dieudonné thought the change unnecessary, despite the criticisms of Skolem [Sk], as would Mac Lane. The day was probably carried by Chevalley, who was interested in Gödel's work on AC and GCH, and Cartan, who was against too fixed and narrow an axiomatic base.

Consider these failings of ZC, documented in the papers cited:

(H·16·0) ZC cannot prove that every set has a rank: see Model 13 in section 7 of [M9, §7];

(H·16·1) ZC cannot prove that every set has a transitive closure: see [M9, §12];

(H·16·2) ZC cannot prove that the class of hereditarily finite sets is a set: see [M6];

(H·16·3) Z cannot conveniently handle Gödel's concept of constructibility: see [M7, §4];

(H·16·4) ZC cannot do Shoenfield-style forcing: see [M11].

(H·16·5) ZC is unable to construct the direct limit discussed in [M7, Example 9.32, p.224];

(H·16·6) ZC cannot prove the determinacy of Borel games: see [Sta].

The reader may sense the problem common to the first five, namely, the absence in Z of explicit forms of replacement, even those supported by the Kripke–Platek system KP. But that amount is there in coded form: it is shown in [M7] that if Z is consistent, so is Z + KP; and in Z+ KP, those first five objections melt away; further, adapting Gödel's proof for ZF to proving the consistency of AC relative to that of Z + KP is straightforward; so the declarations of Weil, Dieudonné and Mac Lane that ZC is plenty for their mathematics merely mean that their mathematics makes very little use of the recursion-theoretic side of mathematics. They missed a lot: had they added to their chosen ZC the axioms of KP, their enhanced theory would be no stronger, consistency-wise, and no nearer the large cardinal axioms they dreaded, but would have given, to them and their followers, conceptual access to the beauty and power of post-Gödelian set theory.

With the last two, the problem is that proving the statement concerned necessarily goes beyond the consistency strength of ZC; very much so in the case of Borel determinacy and other assertions explored by Harvey Friedman; and therefore, if ZC is consistent, such proofs cannot exist even in ZC + KP.

H·17 There is a final class of results which are provable in ZC, or even in MAC, but whose proofs would, conceptually, involve a voyage into the world of ZF. Here are three examples, derived from arguments of Gödel, Solovay, Shelah, Raisonnier, H. Friedman and Martin about projective sets of reals:

if every Σ^1_3 set is Lebesgue measurable, then every uncountable Π^1_1 set has a perfect subset;

if every uncountable Π^1_1 set has a perfect subset, then every Σ^1_2 set is Lebesgue measurable;

every Borel game with integer moves is determined provided every Turing-closed such game is.

H·18 COMMENT The pages of *La Tribu* give the impression that Bourbaki finalised plans for a book only when the criticism and energies of the members had reached exhaustion. It is plain that individual members of the group did think, in their different ways, about foundational matters; one feels that a part, at least, of the oppressiveness of Bourbaki comes from the secrecy and anonymity of their activities, with the consequence that no one person would admit to responsibility for the outcome.

We must here leave our scrutiny of *La Tribu* and return to our discussion of the public consequences of Bourbaki's publications; and as we do, we become aware of the change from the sensitivity of individual perceptions to the crudity of collective decisions; much as Rostropovitch declared [Ste, p.249] a brutal entry of the brass in Lutosławski's Cello Concerto to conjure an image of the Central Committee at full strength.

Why use Bourbaki's formalisation?

H·19 Bourbaki were starting up before the dust had settled from Gödel's discoveries; they wanted to steer clear of the problem of incompleteness; so they made what they thought would be practical decisions; but they could have made better ones. And it is that last message that has not yet reached the public: formalised mathematics need not be the dog's dinner that Bourbaki make of it.

The number 4523659424929 in the title of [M8], when inserted into Google, yields numerous hits, many of which are in Chinese or Japanese, and which, I am told, are contributions to an on-line discussion about the possibility of formalised mathematics on a computer, and that the conclusion being reached is that my calculations in [M8] show that automated theorem proving is an impossibility.

But that conclusion, though reinforced by the grotesque length of terms generated in Bourbaki's later editions, seems premature. The review [Got] of [M8] in *Mathematical Reviews* hints at the existence of simpler formalisations than even that of Bourbaki's first edition; but let us be explicit. Suppose we formalise mathematics with two binary relations $=$ and $\in$, propositional connectives $\neg$, $\&$, individual variables $x, \ldots$, the quantifier $\forall$, and a primitive symbol $\lambdabar$ for the class forming operator, with syntax to match, so that $\lambdabar x\mathfrak{A}$ is what is commonly written as $\{x \mid \mathfrak{A}\}$; then the empty set, $\varnothing$, can be defined in six symbols, as $\lambdabar x\neg x = x$ and its singleton, $\{\varnothing\}$, as $\lambdabar x\forall y\neg y \in x$; eight symbols in all, including just one quantifier.

I ask those who would treat mathematics as a formalised text: why use a formalisation that defines the number One not in eight symbols but in 2409875496393137472149767527877436912979508338752092897?

Structuralism: a part but not the whole of mathematics

Mathematics and logic move on. After Hilbert and, in effect, after Bourbaki came Gödel; and after Gödel's work of the thirties—his completeness theorem, his incompleteness theorem and his relative consistency proof for AC and GCH$^\diamond$—came a further major development in foundational ideas, stemming from the Eilenberg–Mac Lane theory of categories. Though there are differences between Bourbaki and the school of Mac Lane, they are closer to each other than either are to the set-theoretic conception of mathematics, and might conveniently be given the blanket label of *structuralists*.

H·20 Cartier again [Sen]:

> "*Most people agree now that you do need general foundations for mathematics, at least if you believe in the unity of mathematics. I believe now that this unity should be organic, while Bourbaki advocated a structural point of view.*"

$\diamond$ For which, Bernays [Ber2] suggests, the inspiration may have been Hilbert's attempt in [Hi2] to prove CH by enumerating definitions.

"*In accordance with Hilbert's views, set theory was thought by Bourbaki to provide that badly needed general framework. If you need some logical foundations, categories are a more flexible tool than set theory. The point is that categories offer both a general philosophical foundation — that is, the encyclopædic or taxonomic part — and a very efficient mathematical tool to be used in mathematical situations. That set theory and structures are, by contrast, more rigid can be seen by reading the final chapter in Bourbaki's Set Theory, with a monstrous endeavour to formulate categories without categories.*"

In the second quotation it is plain that what Cartier means by set theory is the very limited contents of the Bourbaki volume of that name; a far cry from what set theorists mean by set theory. On the other hand, in the first one, Cartier may be echoing a point made in [M4], that unity is desirable but not uniformity.

H·21 In mentioning uniformity we touch on very dangerous topics; Solzhenitsyn wrote that Stalin made people second-rate; a comment from Feyerabend [Fe] is here relevant:

p 306: "*It is not the interference of the state that is objectionable in the Lysenko case, but the totalitarian interference that kills the opponent instead of letting him go his own way.*"

Even in less threatening circumstances uniformity leads to a failure of critical understanding, and I fear that something of that kind has happened in mathematics as a result of the excessive influence of Bourbaki. Once a mistake is embedded in a monolithic system, it is hard to remove.□

A possible and regrettable consequence of the uniformising tendency of Bourbaki is Grothendieck's withdrawal from the group and subsequently from contact with other mathematicians: one wonders whether he felt threatened by Bourbaki in the same totalitarian way that Chevalley may have been shouted down by Dieudonné, Cantor was blocked by Kronecker, and Nikolai Lusin was menaced by, and Giordano Bruno[54] actually suffered, a death sentence.

H·22 On the question of the unity of mathematics, I should stress that I see the advent of structuralism in mathematics as a bifurcation from, not a development of, set theory. In some areas of mathematics equality

---

□ The mathematical micro-society might here be suffering damage similar to that, analyzed by Hayek in [vHay3], done to macro-society by planning the unplannable.

[54] Links between the conceptions of infinity of Nicholas of Cusa, Bruno and Cantor are discussed in the thoughtful essay of Hauser [Hau].

up to an isomorphism is good enough whereas in others it is not, and structuralism and set theory are on opposite sides of this divide. So while in certain areas of mathematics, structuralism has had a great success, I believe that were the foundational ideas of the structuralists to extinguish all other ideas about the foundations of mathematics, it would be to the great impoverishment of mathematics.[55] Equally, mathematics would be the poorer were set-theoretic ideas to extinguish structuralist ideas: and in [M13] I shall develop the approach outlined in [M12] to probe the nature of the above divide and argue for a pluralist account of the foundations of mathematics. Dieudonné once wrote that we have not begun to understand the relationship between combinatorics and geometry; and I shall hope in [M13] to show that as in a classical tragedy, the Bourbachistes do not realise that what they seek is already to hand.

H·23 Meanwhile something of the dual nature of mathematics is conveyed by the friendly exchange [M]* and [M4] between Mac Lane and myself; and the technically-minded will find in [M7] a close scrutiny of the system of set theory—a subsystem of ZC—that seemed natural to Mac Lane, and in [M9] an even closer scrutiny, if possible, of certain systems of set theory that are important to set theorists.

Mac Lane did not, be it noted, see the issue between topos theory and set theory as an ontological competition. My own view of set theory is that I think of particular superstructures of abstract ideas being called into being to solve particular problems, different superstructures being invoked at different times. I would not be perturbed if the superstructure that solves one problem is not right for another.

Personally I define set theory as the study of well-foundedness, and regard its foundational successes as occurring when it meets a need for a new framework for a "recursive" construction (in a suitably abstract sense). I don't think it succeeds at all in accounting for geometric intuition. That failure should not be allowed to obscure its successes; but nor should its successes be judged a reason for sweeping its failures under the foundational carpet.

There remains the eternal challenge of conveying to others the limitations they are putting on their conceptual universe by adopting exclusively one mode of thought. How does one prove to someone that he is colour-blind? The victim has to be willing to notice that others have perceptions denied to him.**

---

[55] and of economics, as discussed in [M12].

* Is its title an allusion to the "supreme insult" of *La Tribu*, N° 15, page 5?

** A friendly critic of an earlier draft supplies the example of green and red tomatoes, which a red-green colour-blind person can distinguish by taste or by

H·24 But, in mathematics at least, there are signs of these different perceptions. Cartier writes in [Sen]:

> "Following the collapse of the Soviet Union, the Russians
> have brought a different style to the West, a different
> way of looking at the problems, a new blood."

The group centred around Baire, E. Borel, and Lebesgue created a new view of analysis growing out of the insights of Cantor. Both Lusin and Janiszewski came from the East to sit at their feet, and returned home with a positive message. I wonder to what extent the Russian style that Cartier has noticed descends through Lusin from Baire. The writings of Graham and Kantor [GrK1, GrK2, GrK3] are here highly relevant, developing in detail, as they do, topics only touched on in [Sen] and [M1].

Another collapse

H·25 There is a collapse of intellectual level in French schools: see [Coi]. So far as mathematics goes, I believe that the cause of the collapse of mathematical understanding is the suppression of the teaching of logic, which, I suggest, is the consequence of Bourbaki's disastrous treatment of logic.

H·26 ASIDE I am not saying that nothing but logic should be taught, far from it, any more than I should say that an aspiring pianist should play nothing but scales. But training in logic will stand you in good stead in many fields, just as working at the studies of Liszt will give you greater command of the works of Beethoven and Chopin. Logic strengthens the mind just as Liszt strengthens the fingers; and both strengthenings then permit you to go on to greater things.

H·27 For further evidence, let us explore another story of intellectual collapse, namely that of the Italian school of geometry. Mumford writes[56] that the three leaders were Castelnuovo, Enriques and Severi; and that Castelnuovo was totally rigorous, whereas Enriques gave incomplete proofs but was aware of the gaps and tried to fill them; and Severi, after a brilliant start, wrote rubbish; and in effect killed the whole school.

Mumford thinks that the collapse started around 1930. The Italians were not short of ideas, but no one knew what had been proved. Zariski and Weil sorted out the mess; then Grothendieck revolutionised the subject. With Zariski were associated the Bourbachistes Weil, Chevalley, and Samuel.

---

feel but not by sight alone, unlike fully-sighted persons.

[56] Email to Thomas Forster of 23.xi.94, visible at
http://ftp.mcs.anl.gov/pub/qed/archive/209

H·28 Peano died in 1932 and was teaching until the day before his death. Bourbaki listed in their historical note these co-workers of Peano: Vailati, Pieri, Padoa, Vacca, Vivanti, Fano and Burali-Forti. Peano's biographer, Hubert Kennedy, comments that most of them were kept out of university life:

Vailati (1863 - 1909) read engineering at Turin 1880-84. He came under Peano's influence and then read for a mathematics degree which he got in 1888. He went home to Lode, then was assistant to Peano (1892) and then to Vito Volterra. He resigned his university position as assistant in 1899, and became a high school teacher. He then worked on logic and philosophy.

Pieri (1860-1913) graduated at Pisa in 1884, then taught at the Military Academy in Turin. He obtained a doctorate from Turin in 1891, and then taught projective geometry courses there. He got a job in 1900 at the University of Catania in Sicily, then in 1908 moved to Parma.

Padoa (1868-1937) taught in secondary schools and at a Technical Institute in Genoa. He applied unsuccessfully for university lectureships in 1901, 1909 and 1912. Late in life he held a lectureship in mathematical logic at the University of Genoa 1932-1936.

Vacca (1872-1953) was left-wing in politics in his youth; graduated in mathematics from Genoa in 1897; moved to Turin and became Peano's assistant. He discovered the importance of the unpublished works of Leibniz and told Couturat about them. He later took up Chinese language and literature.

I have been unable to discover anything about Vincenzo Vivanti.

Fano (1871-1952) had a rich father. He worked under Felix Klein, and then in Rome (1894) Messina (1899) and from 1901 as Professor at the university of Turin. He was expelled in 1938 by the Fascists. He worked mainly on projective and algebraic geometry.

Cesare Burali-Forti (1861-1931) graduated from Pisa in 1884; from 1887 he taught at the military academy in Turin. He gave an informal series of lectures on mathematical logic at the University of Turin in 1893/4 and was Peano's assistant 1894-6.

H·29 After the second world war, logical studies in Italy revived with the 1948 translation of and commentary on Frege by Ludovico Geymonat (1908–1991), and furthered by Ettore Casari, (1933–) who learnt about the work in Vienna and Germany from Geymonat at Pavia and then completed his studies at Münster, where Heinrich Scholz had built up a school of logic. In the collection [GP] he asks:

"When - in the light of what later occurred - we look at Peano's and his followers' metalogical achievements, a crucial question arises:

how could it be that these skills, these competences, not many years after what Hans Freudenthal liked to call "the Parisian triumph of the Italian phalanx" should have ceased not only to be a reference point for world-wide research, but even to appear on the Italian cultural scene?"

H·30 PROBLEM Is it possible that the collapse of Italian geometry was brought on by the suppression of Italian logic? The dates fit uncomfortably well.

Back to St Benedict

H·31 We must wind up our discussion. So far as Bourbaki's treatment of logic is concerned, the picture is rather sad: Hilbert attacks an admittedly difficult problem, entrusts the work to younger colleagues, and can, perhaps from embarrassment, barely bring himself to acknowledge the Gödel revolution. Bourbaki copy Hilbert's pre-Gödelian position, with its belief that foundational problems can be settled "once and for all", its identification of consistency, truth and provability, and its attempt to declare mathematics to be an uninterpreted calculus; and others copy Bourbaki. Progressively the misunderstanding spreads.

H·32 Eilenberg acknowledged that Bourbaki hadn't thought through their position on foundations clearly and that what they had provided was a mess. But then he felt foundations was always a work in progress, an outlook shared by several members of Bourbaki: but when, later, after the impact of category theory had become evident—especially after the adjoint functor theorem and Grothendieck's work—it was suggested they go back and do a fresh treatment of foundations from scratch, Weil vetoed the idea as a mis-application of energy and resources. Thus Weil, the tyrant, imposed his static view of logic on his colleagues.

It is striking that Chevalley who (as did Cartier) exerted himself to read Gödel's monograph on AC and GCH, spoke out against the boorish conduct of "an ex-member of Bourbaki" in [Ch].

H·33 A distinguished, non-French, mathematician, on reading an earlier draft of this essay, wrote that whilst he had noticed that Bourbaki's logic is very bad, and whilst he acknowledges that I have carefully explained how and why it is bad, nevertheless he does not understand how it is possible not to see the great unifying force and amplitude of the rest of Bourbaki's work. In his view,

"Bourbaki's epoch is gone, but it was a great epoch, and their achievements are as undying as Euclid's. We go forward starting where they ended."

Those comments summarise our dilemma. Bourbaki's structuralist conception advanced certain areas of mathematics but tended to stifle others. How might we undo the admitted harm Bourbaki have done to the understanding and teaching of set theory and logic—subjects not at the forefront of their thought—whilst retaining the benefit of their work in the areas in which they were very much involved?

H·34 That is a non-trivial tactical problem, and I have little faith in the ability of a central committee to solve it by decree. Indeed this whole lamentable saga calls into question the intellectual adequacy of "modern" managerialist centralising universities. It would be better, as I suggest below, to have many independent scholarly bodies such as are almost called for in Article 101 of Napoleon's university statutes.

H·35 One aspect of this problem, for votaries of mathematical logic, is the challenge of imbuing mathematicians with a lively post-Gödelian sense of the vitality of logic. Two encouraging signs for France are the pleasing, if ironical, circumstance that despite Bourbaki's dead hand, Paris has now acquired one of the largest concentrations of logicians on the planet, and the fact that since 2000, of the Sacks prizes bestowed by the Association for Symbolic Logic on doctoral dissertations of outstanding quality written on topics in logic, four have gone to dissertations written at French universities. It is greatly to be hoped that in consequence the trickle-down phenomenon will, over the next twenty years, work in the reverse direction, to restore the teaching of logic to schoolchildren in France and elsewhere.

Such a change is much needed, for on the educational front, a new dark age approaches. Following the dropping of logic in the curriculum, schoolchildren in France are no longer taught to prove theorems; they are given theorems as statements and then given exercises in their application. A generation is growing up without the urge towards rigour. When the battery of their calculator runs down and the calculator starts to make mistakes, how will they know?

H·36 The phenomenon of creativity being arrested by excessive bureaucratic control is well-known to historians of past cultures. I quote from *The Fatal Conceit* by F. A. von Hayek [vHay3] for the following information and references concerning Ancient Egypt, Byzantium and mediæval China.

p 33 *In his study of Egyptian institutions and private law, Jacques Pirenne describes the essentially individualistic character of the law at the end of the third dynasty, when property was 'individual and inviolable, depending wholly on the proprietor' but records the beginning of its decay already during the fifth dynasty.*

Pirenne, J. (1934) *Histoire des Institutions et du droit privé de l'ancienne Egypte* (Brussels: Edition de la Fondation Egyptologique Reine Elisabeth)

This led to the state socialism of the eighteenth dynasty described in another French work of the same date which prevailed for the next two thousand years and largely explains the stagnant character of Egyptian civilization during that period.

> Dairanes, Serge (1934) *Un Socialisme d'Etat quinze Siècles avant Jesus-Christ* (Paris: Libraire Orientaliste P.Geuthner)

p 44 It would seem as if, over and over again, powerful governments so badly damaged spontaneous improvement that the process of cultural evolution was brought to an early demise. The Byzantine government of the East Roman Empire may be one instance of this.

> Rostovtzeff M. (1930) 'The Decline of the Ancient World and its Economic Explanation', *Economic History Review, II*; *A history of the Ancient World* (Oxford: Clarendon Press); *L'empereur Tibère et le culte impérial* (Paris: F.Alcan), and *Gesellschaft und Wirtschaft im Römischen Kaiserreich* (Leipzig: Quelle & Meyer).
>
> Einaudi, Luigi (1948) 'Greatness and Decline of planned economy in the Hellenic world', *Kyklos* II, pp 193–210, 289–316.

p 45 And the history of China provides many instances of government attempts to enforce so perfect an order that innovation became impossible

> Needham, Joseph (1954–85) *Science and Civilisation in China* (Cambridge: Cambridge University Press).

p 46 What led the greatly advanced civilisation of China to fall behind Europe was its governments' clamping down so tightly as to leave no room for new developments, while Europe probably owes its extraordinary expansion in the Middle Ages to its political anarchy.

> Baechler, Jean (1975) *The origin of capitalism* (Oxford: Blackwell), page 77.

H·37 The general challenge to our mathematical culture might be seen as that of regeneration, such as, Charlemagne thought, faced Europe after barbarism engulfed Roman civilisation. But the "green shoots of recovery" spread only slowly from a centre: though the Institut Henri Poincaré might have had logic seminars in 1956, even if not part of a degree course, the Rue d'Ulm began teaching logic only 33 years later. So, like Charlemagne, let us turn for a solution to the decentralist Benedictine idea, the nature and strength of which is well-summarised in a passage [T-R, page 121] of a book by a former Master of Peterhouse:

> "In the darkening, defensive days of the sixth century, the Benedictine monastery had been the cell of Christendom: every cell independent, so that if one cell failed, another might survive."

That independence still holds today: though they all follow the Rule of St Benedict, the Benedictine abbeys do not form an order; each is independent and each abbot is sovereign. The Benedictine idea led, in Oxford and Cambridge, to the founding of colleges, the founding statutes of the oldest colleges being expressly based on the Rule of St Benedict. The strength of Oxford and Cambridge as universities derives from the traditional independence of each college within the university, exemplified by the comment of an earlier Master of Peterhouse, the mathematician Charles Burkill, on a University proposal to establish a centralised Needs Committee for the totality of colleges, that "such a committee can only be mischievous".

There is no reason why the Oxbridge system of colleges, that is, of self-governing, self-recruiting, property-owning communities of scholars, should not be permitted to develop in other countries, and thereby encourage the "free market" approach to education expressed in my final quotation, from Feyerabend's most famous book, *Against Method*:

p 30 *"[Knowledge] is not a gradual approach to the truth. It is rather an ever increasing ocean of mutually incompatible and perhaps even incommensurable alternatives, each single theory, each fairy tale, each myth that is part of the collection forcing the others into greater articulation, and all of them contributing via this process of competition to the development of our consciousness."*

Acknowledgments

The author is much indebted to those friends, mathematicians, logicians, philosophers and historians who proffered encouraging, helpful, informative and in many cases detailed comments on earlier versions: Joan Bagaria, Liliane Beaulieu, John Bell, Pierre Cartier, Carles Casacuberta, Leo Corry, Bruno Courcelle, James Cummings, Kenneth Derus, Jean-Jacques Duby, Olivier Esser, David Fremlin, Thomas Forster, Jason Fordham, Marcel Guillaume, Labib Haddad, Kai Hauser, Daisuke Ikegami, Akihiro Kanamori, Jean-Michel Kantor, Otto Kegel, Juliette Kennedy, Wendula Gräfin von Klinckowstroem, Georg Kreisel, Kenneth Kunen, Nicholas de Lange, Menachem Magidor, Yuri Manin, David McKie, Colin McLarty, William Messing, Heike Mildenberger, Keith Mitchell, Ieke Moerdijk, Charles Morgan, Marianne Morillon, Pierre-Eric Mounier-Kuhn, Sir Edward Mountain B^t, Anil Nerode, Maurice Pouzet, J. Barkley Rosser J^r, Gerald Sacks, Wilfried Sieg, Basil Smith, Jacques Stern, David Thomas, Dominique Tournès, Jouko Väänänen, Giorgio Venturi, Victor Walne, Philip Welch and Michael Wright.

References

[Ack1] W. Ackermann, Begründung des "tertium non datur" mittels der Hilbertscher Theorie der Widerspruchsfreiheit. *Mathematische Annalen* **93** (1924) 1–36.

[Ack2] W. Ackermann, Zur Widerspruchsfreiheit der Zahlentheorie, *Mathematische Annalen* **117** (1940) 162–194.

[Ad] J. F. Adams, *Stable Homotopy and Generalised Homology*, Chicago Lectures in Mathematics, 1974.

[AF] J. F. Adams, *Localisation and Completion*, Chicago lecture notes taken by Zig Fiedorowicz in Spring 1973 and revised and supplemented by him in 2010.

[Ag] Pierre Ageron, La philosophie mathématique de Roger Apéry, *Philosophia Scientiæ*, Cahier Spécial 5, «*Fonder autrement les mathématiques*» (2005) 233–256.

[An] Irving H. Anellis, Peirce rustled, Russell pierced, *Modern Logic* **5** (1995), 270–328.

[Ap1] François Apéry, *Roger Apéry, 1916-1994: A Radical Mathematician*, *Mathematical Intelligencer* **18** (2) (1996) 54–61.

[Ap2] François Apéry, *Un mathématicien radical*, Mulhouse: autoédition, 1998, 176 pp.

[BCM] J. Bagaria, C. Casacuberta and A. R. D. Mathias, Epireflections and supercompact cardinals, *Journal of Pure and Applied Algebra* **213** (2009) 1208–1215.

[BCMR] J. Bagaria, C. Casacuberta, A. R. D. Mathias and Jiří Rosický, Definable orthogonality classes are small, http://arxiv.org/abs/1101.2792

[Bea1] L. Beaulieu, Bourbaki's art of memory, *Osiris* **14** (1999) 219–251.

[Bea2] L. Beaulieu, *Bourbaki: History and Legend, 1934-1956*, Springer-Verlag, 2006.

[Ber1] Paul Bernays, Axiomatische Untersuchungen des Aussagen-Kalkuls der "Principia Mathematica", *Mathematische Zeitschrift* **25** (1926) 305–320.

[Ber2] Paul Bernays, review of Gödel's book on GCH, *Journal of Symbolic Logic* **5** (1940) 117–118.

[Bl1] M. Black, review in *Mind* **49** (1940) 239–248.

[Bl2] M. Black, review in *Journal of Symbolic Logic* **13** (1948) 146–7.

[Bor] A. Borel, Twenty-five years with Nicolas Bourbaki, 1949–1973, *Notices of the American Mathematical Society*, 1998, 373–380.

[Bou49] N. Bourbaki, Foundations of Mathematics for the Working Mathematician, *Journal of Symbolic Logic* **14** (1949) 1–8.

[Bou54] N. Bourbaki, *Théorie des Ensembles,* 1st edition: Hermann, Paris. Chapters 1, 2, 1954; Chapter 3, 1956; Chapter 4, 1957.

[Bous1] A. K. Bousfield, The localization of spaces with respect to homology, *Topology* **14** (1975), 133–150.

[Bous2] A. K. Bousfield, The localization of spectra with respect to homology, *Topology* **18** (1979), 257–281.

[Bu] Stanley Burris, Notes on [HiA], March 13, 2001, `http://www.math.` `uwaterloo.ca/~snburris`

[CaScS] C. Casacuberta, D. Scevenels and J. H. Smith, Implications of large-cardinal principles in homotopical localization, *Advances in Mathematics* **197**, (2005), 120-139.

[Ch] C. Chevalley *et. al.*, Pour la liberté en mathématiques, *Gazette des mathématiciens* **18** (1982) 71–74.

[Cho] M. Chouchan, *Nicolas Bourbaki, Faits et Légendes,* Editions du Choix, Argenteuil, 1995. A copy is in the library at Göttingen, reference FMAG 98 A 8296. ISBN 2.909028-18-6.

[Chu] Alonzo Church, review of [De], *Journal of Symbolic Logic* **13** (1948) 144.

[Coh1] P. Cohen, The independence of the continuum hypothesis, *Proc. Nat. Acad. of Sci. U.S.A.*, **L** (1963) 1143–48, **LI** (1964) 105–10.

[Coh2] P. Cohen, *Set Theory and the Continuum Hypothesis,* Addison-Wesley, 1966.

[Coi] S. Coignard, *Le pacte immoral: comment ils sacrifient l'éducation de nos enfants,* Albin Michel, 2011.

[Cor1] L. Corry, Nicolas Bourbaki and the concept of mathematical structure, *Synthese* **92** (1992) 315–348.

[Cor2] L. Corry, *Modern Algebra and the Rise of Mathematical Structures,* Science Networks Historical Studies **17**, Birkhäuser-Verlag, Basel, Boston, Berlin, 1996.

[Cor3] L. Corry, The Origins of Eternal Truth in Modern Mathematics: Hilbert to Bourbaki and Beyond; *Science in Context* **10** (1997) 253–296.

[Cor4] L. Corry, *David Hilbert and the Axiomatization of Physics (1898–1918),* Dordrecht: Kluwer, 2004.

[Cor5] L. Corry, Axiomatics, Empiricism, and Anschauung in Hilbert's Conception of Geometry: Between Arithmetic and General Relativity, in *The Architecture of Modern Mathematics: Essays in History and Philosophy,* edited by Jeremy Gray and José Ferreirós, pages 155-176. Oxford University Press (2006).

[Cu] Haskell B. Curry, review of Rosser [Ro2] *Bull. Amer. Math. Soc.* **60** (1954) 266–272.

[CW] *Correspondance entre Henri Cartan et André Weil (1928-1991)*, edited by Michèle Audin, Documents Mathématiques de France, tome 6, Société Mathématique de France, 2011.

[vDa1] D. van Dalen, The War of the Frogs and the Mice, *The Mathematical Intelligencer* **12** (1990) 17–31.

[vDa2] D. van Dalen, *Mystic, Geometer and Intuitionist: The Life of L. E. J. Brouwer*, Oxford University Press, in two volumes: *The dawning Revolution* (1999), *Hope and Disillusion* (2005).

[Daw] John W. Dawson, Jr., The reception of Gödel's incompleteness theorems, *Proceedings of the Biennial Meeting of the Philosophy of Science Association*, (1984) 253–271.

[De] A. Denjoy, *L'énumeration transfinie. Livre I. La notion de rang*, Gauthier-Villars, Paris, 1944.

[Di1] Jean Dieudonné, *Foundations of Modern Analysis*, Academic Press, New York, 1960.

[Di2] Jean Dieudonné, *Choix d'Œuvres de Jean Dieudonné de l'Institut*, Hermann, Paris, 1981, Tome I.

[Di3] Jean Dieudonné, *Pour l'honneur de l'esprit humain*, Hachette 1987; Englished as *Mathematics: the music of reason*, Springer-Verlag, 1992.

[DrKa] B. Dreben and A. Kanamori, Hilbert and Set Theory, *Synthèse*, **110**, (1997) 77–125.

[Du] T. Düppe, http://www.econ-pol.unisi.it/eche07/PaperDuppe.pdf or, expanded, as *Part III of* http://publishing.eur.nl/ir/repub/asset/16075/Proefschrift%20Till%20Duppe%5Blr%5D.pdf

[ES] W.B. Ewald and W. Sieg (eds), *David Hilbert's lectures on the foundations of mathematics and physics, 1891-1933*, volume 3; Springer Verlag, to appear in 2014.

[Fe] Paul Feyerabend, *Against Method*, New Left Books, 1975.

[Fr-bH-L] A. A. Fraenkel, Y. bar-Hillel, A. Lévy, *Foundations of Set Theory*, North Holland Studies in Logic, volume 67, second revised edition, 1973.

[G1] R. O. Gandy, review of [Bou54], Chapters 1 and 2, *Journal of Symbolic Logic* **24** (1959) 71–73.

[G2] Gerhard Gentzen, Die Widerspruchsfreiheit der reinen Zahlentheorie, *Math. Ann.* **112** (1936) 492–565.

[G3] Gerhard Gentzen, *Die gegenwartige Lage in der mathematischen Grundlagenforschung. Neue Fassung des Widerspruchsfreiheitsbeweises für die reine Zahlentheorie*, Forschungen zur Logik &c, Neue Folge, Heft 4. Leipzig: S. Hirzel, 1938.

[Gd] R. Godement, *Cours d'Algèbre*, Hermann, Paris 1963; Englished as *Algebra*, Kershaw, London, 1969.

[Gö] Kurt Gödel, *Collected Works*, edited by S. Feferman and J. W. Dawson, in five volumes; Oxford University Press, 1986, 1990, 1995, 2003, 2003.

[Gol] W. Goldfarb, Logic in the twenties: the nature of the quantifier, *Journal of Symbolic Logic* **44** (1979) 351–368.

[Got] Siegfried Gottwald, Review of [M8], *Mathematical Reviews* (2004a:03009)

[GP] *Giuseppe Peano between mathematics and logic*, edited by Fulvia Skof, Springer Milan New York 2011.

[GrK1] L. Graham and J.-M. Kantor, A Comparison of two cultural approaches to Mathematics: France and Russia, 1890-1930, in *Isis*, March 2006.

[GrK2] L. Graham and J.-M. Kantor, Religious Heresy and Mathematical Creativity in Russia, *Mathematical Intelligencer*, **29** (4), (2007) 17–22.

[GrK3] L. Graham and J.-M. Kantor, *Naming Infinity: a true story of religious mysticism and mathematical creativity*, Harvard University Press, March 2009.

[Had] J. Hadamard and others, Cinq Lettres, *Bulletin de la Société Mathématique de France* **33** (1905) 261–273.

[Hau] Kai Hauser, Cantor's Absolute in Metaphysics and Mathematics, *International Philosophical Quarterly* **53** (2013) 161–188

[vHay1] Friedrich August von Hayek, The Pretence of Knowledge, *Nobel Memorial Prize Lecture, 1974*, at `http://nobelprize.org/nobel_prizes/economics/laureates/1974/hayek-lecture.html`

[vHay2] Friedrich A. von Hayek, Die Anmaßung von Wissen, in F.A. von Hayek, (Gesammelte Schriften in deutscher Sprache, Abteilung A Band 1,) *Wirtschaftstheorie und Wissen, Aufsätze zur Erkenntnis- und Wissenschaftslehre*, herausgegeben von Viktor Vanberg, 87–98. Tübingen: Mohr Siebeck, 2007.

[vHay3] Friedrich A. von Hayek, *The Fatal Conceit: the Errors of Socialism*, Routledge, London, 1988.

[vHe] Jean van Heijenoort, (editor) *From Frege to Gödel: a Source Book in Mathematical Logic, 1879–1931*. Harvard University Press, 1967.

[Hi1] D. Hilbert, Die logischen Grundlagen der Mathematik, *Mathematische Annalen* **88** (1923) 151–165.

[Hi2] D. Hilbert, Über das unendliche, *Mathematische Annalen* **95** (1925) 161–190; translated by André Weil as *Sur l'infinie*, *Acta Mathematica* **48** (1926) 91–122.

[Hi3] D. Hilbert, Die Grundlagen der Mathematik, *Abhandlungen aus dem mathematischen Seminar der Hamburgischen Universität* **6** (1928) 65–85.

[Hi4] D. Hilbert, Probleme der Grundlegung der Mathematik, *Mathematische Annalen* **102** (1930) 1–9.

[Hi5] D. Hilbert, Die Grundlegung der elementaren Zahlenlehre, *Mathematische Annalen* **104** (1931) 485–494.

[Hi6] D. Hilbert, Beweis des *tertium non datur*, *Nachrichten von der Gesellschaft der Wissenschaften zu Göttingen, Mathematisch-physicalische Klasse* (1931) 120–125.

[HiA] D. Hilbert and W. Ackermann, *Grundzüge der theoretischen Logik*, Springer-Verlag, Berlin, 1928.

[HiB1] D. Hilbert and P. Bernays, *Grundlagen der Mathematik, Band 1*, Springer-Verlag, Berlin, 1934.

[HiB2] D. Hilbert and P. Bernays, *Grundlagen der Mathematik, Band 2*, Springer-Verlag, Berlin, 1939.

[Hu] Edward V. Huntington, The interdeducibility of the new Hilbert–Bernays theory and Principia Mathematica, *Annals of Mathematics* **36** (1935) 313–324.

[Ka] Akihiro Kanamori, Cohen and set theory, *Bulletin of Symbolic Logic* **14** (2008) 351–378.

[Ke] Juliette Kennedy, Gödel's thesis: an appreciation, www.amazon.com/ Kurt-Gödel-Foundations-Mathematics-Horizons/dp/0521761441

[Kl] S. C. Kleene, *Introduction to Metamathematics*, Princeton, van Nostrand, North Holland, 1952.

[L] A. C. Leisenring, *Mathematical Logic and Hilbert's Epsilon Symbol*, London, 1969.

[L-F,A] J. Lelong-Ferrand & J.-M. Arnaudiès, *Algèbre*, Editions Dunod, 1978; third edition, 1995.

[Lu] Jan Łukasiewicz, *Aristotle's Syllogistic from the Standpoint of Modern Formal Logic*, second edition. Oxford: Clarendon Press, 1957.

[M] S. Mac Lane, Is Mathias an ontologist? in *Set Theory of the Continuum*, edited by H. Judah, W. Just, and H. Woodin, pp 119–122. MSRI Publications Volume 26, Springer-Verlag, 1992. MR 94g:03011

[Mas1] M. Mashaal, *Bourbaki: une société sécrète de mathématiciens*, Pour la Science, 2002.

[Mas2] M. Mashaal, *Bourbaki: a secret society of mathematicians*, American Mathematical Society, 2006.

[M1] A. R. D. Mathias, The real line and the universe, *Logic Colloquium '76*, edited by R. O. Gandy and J. M. E. Hyland, *Studies in Logic* **87** (North Holland), 531–546; MR 57 # 12218

[M2] A. R. D. Mathias, Logic and Terror, in *Jahrb. Kurt-Gödel-Ges.* (1990) 117–130 MR 92m:01039, and also, in a longer version, in *Physis Riv. Internaz. Storia Sci (N.S.)* **28** (1991) 557–578; MR 93d:03014

[M3] A. R. D. Mathias, The Ignorance of Bourbaki, in *Mathematical Intelligencer* **14** (1992) 4–13 MR 94a: 03004b, and also in *Physis Riv. Internaz. Storia Sci (N.S.)* **28** (1991) 887–904; MR 94a:03004a; translated by Racz András as *Bourbaki tévútjai*, in *A Természet Világa*, 1998, III. különszáma; and by José Maria Almira Picazo as *La ignorancia de Bourbaki*, in *La Gaceta de la Real Sociedad Matemàtica Española*, 7, (2004) 727–748.

[M4] A. R. D. Mathias, What is Mac Lane missing? in *Set Theory of the Continuum*, edited by H. Judah, W.Just, H.Woodin, pp 113-118, MSRI Publications Volume 26, Springer-Verlag, 1992. MR 94g:03010

[M5] A. R. D. Mathias, Strong statements of analysis, *Bull. London Math. Soc.* **32** (2000) 513–526. MR 2001h:03102a, b.

[M6] A. R. D. Mathias, Slim models of Zermelo Set Theory, *Journal of Symbolic Logic* **66** (2001) 487–496. MR 2003a:03076

[M7] A. R. D. Mathias, The Strength of Mac Lane Set Theory, *Annals of Pure and Applied Logic,* **110** (2001) 107–234; MR 2002g:03105.

[M8] A. R. D. Mathias, A term of length 4,523,659,424,929, *Synthese* **133** (2002) 75–86.

[M9] A. R. D. Mathias, Weak systems of Gandy, Jensen and Devlin, in *Set Theory: Centre de Recerca Matemàtica, Barcelona 2003-4*, edited by Joan Bagaria and Stevo Todorcevic, pages 149–224, Trends in Mathematics, Birkhäuser Verlag, Basel, 2006.

[M10] A. R. D. Mathias, Unordered pairs in the set theory of Bourbaki 1949, *Archiv der Mathematik* **94**, (2010) 1–10.

[M11] A. R. D. Mathias, Set forcing over models of Zermelo or Mac Lane, in *One hundred years of axiomatic set theory*, edited by R. Hinnion and T. Libert, Cahiers du Centre de logique, Vol. 17, pages 41–66. Academia-Bruylant, Louvain-la-Neuve, 2010.

[M12] A. R. D. Mathias Economics, common sense, logic and mathematics: are they related? *in preparation*

[M13] A. R. D. Mathias, Endometics and aëxitrophics. *in preparation.*

[Mi] O. H. Mitchell, On a new algebra of logic. In *Studies in logic* by members of the Johns Hopkins University, edited by C. S. Peirce, 72–106. Boston 1883: Little & Brown.

[M-K] Pierre Mounier-Kuhn, *L'informatique en France, de la Seconde Guerre mondiale au Plan Calcul: l'Emergence d'une science*, Presses de l'Université Paris-Sorbonne, février 2010, 718 pp.

[vN] J. v. Neumann, Zur Hilbertschen Beweistheorie, *Mathematische Zeitschrift* **26** (1927) 1–46.

[OH] Claire Ortiz Hill, Frege's Letters, in: *From Dedekind to Gödel*, edited by J. Hintikka, Dordrecht, Kluwer 1995, pp 97–118. Also available at `http://rancho.pancho.pagesperso-orange.fr/Letters.htm`

[Pec] Volker Peckhaus, Paradoxes in Göttingen, in: *One Hundred Years of Russell's Paradox. Mathematics, Logic, Philosophy*, edited by Godehard Link, de Gruyter: Berlin, New York 2004, pages 501-515. (de Gruyter Series in Logic and Its Applications 6)

[Pei] C. S. Peirce, On the Algebra of Logic: a contribution to the philosophy of notation, *American Journal of Mathematics* **7** (1885) 180–96 and 197–202.

[PlS] *Pour la Science*, Bourbaki: une société secrète de mathématiciens, special issue in the series *Les Génies de la Science*, February—May 2000, written by Maurice Mashaal; later expanded as [Mas1] and [Mas2].

[Pu] Hilary Putnam, Peirce the logician, *Historia Mathematica* **9** (1982) 290–301.

[Q] W. v. O. Quine, *The Time of My Life*, M.I.T. Press, Cambridge, Massachusetts, 1985.

[RaTh] B. Rang and W. Thomas, Zermelo's discovery of the "Russell Paradox", *Historia Math.* **8** (1981) 15–22.

[Re1] Constance Reid, *Hilbert*, Springer Verlag, 1970/1996.

[Re2] Miles Reid, *Undergraduate Algebraic Geometry*, LMS Student Texts **12**, Cambridge University Press, 1988, 1994.

[Ro1] J. Barkley Rosser, review of [Bou49], *Journal of Symbolic Logic* **14** (1950) 258–9.

[Ro2] J. Barkley Rosser *Logic for mathematicians*, McGraw–Hill, New York, 1953.

[Ro3] J. Barkley Rosser *Simplified independence proofs*, Academic Press, New York, 1969.

[RoJr] J. Barkley Rosser Jr, On the foundations of mathematical economics, *available at* `http://cob.jmu.edu/rosserjb/`

[Seg1] Sanford L. Segal, *Zentralblatt* review of [M3], Zbl 0764.01009

[Seg2] Sanford L. Segal, *Mathematicians under the Nazis*, Princeton University Press, 2003.

[Sen] Marjorie Senechal, The continuing silence of Bourbaki — an interview with Pierre Cartier, June 18, 1997, *Mathematical Intelligencer.* **20** (1998) 22–8.

[Si1] Wilfried Sieg, Hilbert's Programs, 1917-22, *Bulletin of Symbolic Logic* **5** (1999) 1–44.

[Si2] Wilfried Sieg, Only two letters: the correspondance between Herbrand and Gödel, *Bulletin of Symbolic Logic* **11** (2005) 172–184.

[Si3] Wilfried Sieg, Hilbert's Proof Theory, in the *Handbook of the History of Logic*, edited by Dov M. Gabbay and John Woods, Volume 5, "Logic from Russell to Church", Elsevier, 2008.

[Si4] Wilfried Sieg, In the shadow of incompleteness: Hilbert and Gentzen: pp 87–127 of *Epistemology versus Ontology: essays on the philosophy and foundations of mathematics in honour of Per Martin-Löf*, edited by Peter Dybjer, Sten Lindström, Erik Palmgren and Göran Sundholm, Springer Verlag, 2012.

[Sk] Thoralf Skolem, Einige Bemerkungen zur axiomatischen Begründung der Mengenlehre, *Wiss. Vorträge gehalten auf dem 5. Kongress der skandinav. Mathematiker in Helsingfors 1922*, 1923, 217–232.

[Sta] Lee J. Stanley, Borel diagonalization and abstract set theory: recent results of Harvey Friedman, in *Harvey Friedman's research on the foundations of mathematics*, edited by L. A. Harrington, M. D. Morley, A. Ščedrov and S. G. Simpson, North Holland, 1985.

[Ste] Michael Steinberg, *The Concerto*, Oxford University Press, New York, Oxford, 1998.

[T] D. F. Tovey, *A companion to "The Art of Fugue"*, Oxford University Press, 1931.

[T-R] Hugh Trevor-Roper, *The Rise of Christian Europe*, London: Thames and Hudson, 1965, 1966, 1989.

[WR] Alfred North Whitehead and Bertrand Russell, *Principia Mathematica*, volume 1, Cambridge University Press, 1910.

[Za1] Richard Zach, Completeness before Post: Bernays, Hilbert and the development of propositional logic, *Bulletin of Symbolic Logic* **5** (1999) 331–366.

[Za2] Richard Zach, The practice of finitism: epsilon calculus and consistency proofs in Hilbert's program, *Synthese* **137** (2003) 211–259.

[Ze1] Ernst Zermelo, Beweis, dass jede Menge wohlgeordnet werden kann (Aus einem an Herrn Hilbert gerichteten Briefe), *Mathematische Annalen* **59** (1904) 514–516.

[Ze2] Ernst Zermelo, Neuer Beweis für die Möglichkeit einer Wohlordnung, *Mathematische Annalen* **65** (1908) 107–128.

[Ze3] Ernst Zermelo, Untersuchungen über die Grundlagen der Mengenlehre I, *Mathematische Annalen* **65** (1908) 261–281.

TOWARD OBJECTIVITY IN MATHEMATICS

Stephen G. Simpson

Department of Mathematics
Pennsylvania State University
State College, PA 16802, USA
simpson@math.psu.edu
http://www.math.psu.edu/simpson/

We present some ideas in furtherance of objectivity in mathematics. We call for closer integration of mathematics with the rest of human knowledge. We note some insights which can be drawn from current research programs in the foundations of mathematics.

1. Objectivity and Objectivism

I am a mathematician, not a philosopher. However, as a mathematician and a human being, I have always had the greatest respect for philosophy, and I have always recognized the need for philosophical guidance.

My thinking is largely informed by a particular philosophical system:

Objectivism (with a capital "O").

A key reference for me is Leonard Peikoff's treatise [4]. By the way, Peikoff obtained his Ph.D. degree in Philosophy here at New York University in 1964. His thesis advisor was Sidney Hook.

For those not familiar with Objectivism, let me say that it is a coherent, integrated, philosophical system which encompasses the five main branches of philosophy: metaphysics, epistemology, ethics, politics, aesthetics.

As the name "Objectivism" suggests, the concept of *objectivity* plays a central role in the system. Because objectivity is an epistemological concept, let me say a little about the Objectivist epistemology. Of course, my brief account of the Objectivist epistemology cannot be fully understood outside the context of certain other aspects of Objectivism which I do not plan to discuss here.

The main point is that Objectivist epistemology calls for a close relationship between *existence* (the reality which is "out there") and *consciousness* (a volitional process that takes place within the human mind).

(1) According to Objectivism, *knowledge* (i.e., human, conceptual knowledge) is "grasp of an object by means of an active, reality-based process which is chosen by the subject."

(2) According to Objectivism, *objectivity* is a specific kind of relationship between reality ("out there") and consciousness ("in here").

(3) All knowledge is *contextual*, i.e., it must be understood within a context. Moreover, the ultimate context is the totality of human knowledge. Thererore, all of human knowledge must be integrated into a coherent system. Compartmentalization is strongly discouraged (more about this later).

(4) In integrating human knowledge into a coherent whole, the method of integration is *logic*, defined as "the art of non-contradictory identification." Here "identification" refers to the conceptual grasp of an object or entity in reality.

(5) All knowledge is *hierarchical*. Concepts must be justified or validated by reference to earlier concepts, which are based on still earlier concepts, etc., all the way down to the perceptual roots. This validation process is called *reduction*.

We may contrast Objectivism with two other types of philosophy: *intrinsicism* (e.g., Plato, Augustine) and *subjectivism* (e.g., Kant, Dewey).

(1) To their credit, the *intrinsicists* recognize that knowledge must conform to reality. However, intrinsicism goes overboard by denying the active or volitional nature of consciousness. According to intrinsicism, the process of acquiring knowlege is essentially passive. It consists of "revelation" (Judeo-Christian theology) or "remembering" (Plato) or "intuition," not volitional activity. The operative factor is *existence* rather than consciousness.

(2) To their credit, the *subjectivists* recognize that revelation is not a valid means of cognition. However, subjectivism goes too far by insisting that concepts are not based on reality but rather are created solely out of the resources of our own minds. There are several versions of subjectivism. In the personal version, each individual creates his own universe. In the

social or collective version, concepts and facts are created by a group. In all versions of subjectivism, the operative factor is *consciousness* rather than existence.

Objectivism strikes a balance by emphasizing a close relationship between existence and consciousness. Each of these two factors is operative. Their close relationship is summarized in a slogan:

"Existence is identity; consciousness is identification."

2. Mathematics as Part of Human Knowledge

A major problem in universities and in society generally is *compartmentalization*. Compartmentalization is a kind of overspecialization in which one regards one's own specialty as an isolated subject, unrelated to the rest of human knowledge. Thus, the teachings of one university department (e.g., the English department) may flatly contradict those of another (e.g., the business school) and this kind of situation is regarded as normal.

Compartmentalization can sometimes exist within a single individual. An example is the conservative economist who advocates the profit motive in economics and the Sermon on the Mount in church. Another example is the legislator who calls for strict government control of political advocacy and commercial activity, while at the same time paying lip service to freedom of speech and association.

Here I wish to focus on compartmentalization in the university context, with which I am very familiar.

I am a professor of mathematics at a large state university, Penn State. At our main campus in the appropriately named Happy Valley in Pennsylvania, there are more than 40,000 undergraduate students as well as thousands of graduate students and postdocs.

At the Pennsylvania State University as at most other large universities, much of the research activity is mathematical in nature. Mathematics, statistics, and large-scale computer simulations are heavily used as research tools. This applies to the majority of academic divisions of the university: not only physical sciences and engineering, but also biological sciences, agriculture, business, social sciences, earth sciences, materials science, medicine, and even humanities. In addition Penn State has an Applied Research Laboratory which performs classified, defense-related research and has a huge annual budget. There also, mathematics is heavily used.

What is interesting is that our Department of Mathematics is largely uninvolved in this kind of activity. When mathematicians and non-mathematicians try to collaborate, both sides are often frustrated by "communication difficulties" or "failure to find common ground," due largely to lack of a common vocabulary and conceptual framework. To me this widespread frustration suggests a failure of integration.

As an aside, we can see the detrimental effects of a lack of mathematics in public affairs. Basic mathematical and statistical knowledge is astonishingly rare among the voting public. Lack of quantitative understanding of relative benefits and relative risks may stifle innovation. "A trillion is the new billion," and angry mobs with pitchforks may lose sight of the decimal point.

But, back to the university context. From interactions with mathematicians and non-mathematicians at Penn State and elsewhere, I see a need for greater integration of mathematics with the rest of human knowledge. We need to somehow overcome the compartmentalization which isolates mathematics from application areas.

Philosophy is the branch of knowledge that deals with the widest possible abstractions – concepts such as justice, friendship, and objectivity. Therefore, only philosophy can act as the ultimate integrator of human knowledge. A crucial task for philosophers of mathematics is to provide general principles which can guide both mathematicians and users of mathematics.

Some of the most pressing issues involve *mathematical modeling*. By a mathematical model I mean an abstract mathematical structure M (e.g., a system of differential equations) together with a claimed relationship between M and a real-world situation R (e.g., a weather system). Typically, the mathematician designs the structure M, and the non-mathematician decides which assumptions (e.g., initial conditions) are to be fed into M and how to interpret the results in R. Such models are used extensively in engineering, finance, economics, climate studies, etc.

Some currently relevant questions about mathematical modeling are as follows. What are the appropriate uses of quantitative financial models in terms of risk and reward? Would it be ethical to incorporate the prospect of government bailouts into such models? What are appropriate limitations on the role of mathematical modeling in climate studies? Under what circumstances is it ethically appropriate to base public policy on such models? Etc., etc.

By nature such questions are highly interdisciplinary and require a broad perspective. Therefore, it seems reasonable to think that such questions may be a proper object of study for philosophers of mathematics. It would be wonderful if philosophers could provide a valid framework or standard for answering such questions. Of course, it goes without saying that this kind of philosophical activity would have to be based on a coherent philosophical system including an integrated view of human knowledge as a whole and the role of mathematics within it.

3. Set Theory and the Unity of Mathematics

As is well known, mathematicians tend to group themselves into research specialties: analysis, algebra, number theory, geometry, topology, combinatorics, ordinary differential equations, partial differential equations, mathematical logic, etc. Each of these groups holds its own conferences, edits its own journals, writes letters of recommendation for its own members, etc. Furthermore, among these groups there is frequent and occasionally bitter rivalry with respect to academic hiring, research professorships, awards, etc.

As an antidote to this kind of fragmentation, high-level mathematicians frequently express an interest in promoting *the unity of mathematics*. An 11th commandment for mathematicians has been proposed:

"Thou shalt not criticize any branch of mathematics."

A variant reads as follows:

"*All* mathematics is difficult; *all* mathematics is interesting."

Partly as a result of such considerations, research programs which combine several branches of mathematics are highly valued. Examples of such programs are algebraic topology, geometric functional analysis, algebraic geometry, geometric group theory, etc. Such programs are regarded as valuable partly because they draw together two or more research subcommunities within the larger mathematical community.

As regards the unity of mathematics, set theory has made at least one crucial contribution. Namely, the well known formalism of ZFC, Zermelo-Fraenkel set theory (based on classical first-order logic and including the Axiom of Choice) is a huge achievement. The ZFC formalism provides two extremely important benefits for mathematics as a whole: a common framework, and a common standard of rigor.

S. G. Simpson

(1) ZFC provides the orthodox, commonly accepted framework for virtually all of contemporary mathematics. Indeed, advanced undergraduate textbooks in almost all branches of mathematics frequently include either an appendix or an introductory chapter outlining the common set-theoretic notions: sets, functions, union, intersection, Cartesian product, etc.

(2) The ZFC framework is sufficiently simple and elegant so that all mathematicians can easily gain a working knowledge of it. There is only one basic concept: *sets*. The axioms of ZFC consist of easily understood, plausible, self-evident assumptions concerning the universe of sets. Moreover, the ZFC framework is flexible and far-reaching; within it one can easily and quickly construct isomorphs of all familiar mathematical structures including the natural number system, the real number system, Euclidean spaces, manifolds, topological spaces, Hilbert space, operator algebras, etc.

(3) Among mathematicians, there is little or no controversy about what it would mean to rigorously prove a mathematical theorem. All such questions are answered by saying that the proof must be formalizable in ZFC, i.e., deducible from the axioms of ZFC using standard logical axioms and rules. In his talk yesterday, Professor Gaifman gave an admirably detailed description of how this ZFC-based verification process works in practice.

It is noteworthy that similarly clear standards of rigor do not currently exist in other sciences such as physics, economics, or philosophy.

(4) Mathematicians are highly appreciative of the existence of a common framework and standard of rigor such as ZFC provides.

For instance, there is currently little or no controversy surrounding the Axiom of Choice such as took place in the early 20th century. Virtually all mathematicians are happy and relieved to know that this and similar controversies have been laid to rest.

(5) This comfortable situation allows "working mathematicians" to get on with their research, secure in the belief that they will not be undercut by some obscure foundational brouhaha. Mathematicians appreciate ZFC because it seems to relieve them of the need to bother with foundational questions.

On the other hand, mathematicians have some justifiable reservations about set-theoretic foundations. The existence of a variety of models of ZFC (the set-theoretic "multiverse") is somewhat unsettling, at least for those mathematicians who take foundations seriously. Some mathematicians deal with this kind of uncertainty by asserting that questions such as the Continuum Hypothesis and large cardinals are unlikely to impinge on their own branch of mathematics, or at least their own research within that branch. Some mathematicians even make a point of avoiding higher set theory, for fear of running into such scary monsters. (Of course we mathematical logicians know or strongly suspect that they are whistling in the dark.)

Even worse, when we contemplate the philosophical task which was outlined in Section 2 above, the program of set-theoretic foundations based on systems such as ZFC seems unhelpful to say the least. There seems to be no clear path toward integration of set theory with the rest of human knowledge. Infinite sets and the cumulative hierarchy present a stumbling block. It is completely unclear how to reduce a concept such as $\aleph_\omega$ to referents in "the real world out there." We have no idea whatsoever of how to understand the Continuum Hypothesis as a question about "the real world out there." What in "the real world out there" are the set theorists talking about? The answer seems unclear, and nobody can agree on how to proceed.

Thus it emerges that the program of set-theoretic foundations, useful though it has been in promoting the unity of mathematics and defining a standard of mathematical rigor, appears to stand as an obstacle in the way of a highly desirable unification of mathematics with the rest of human knowledge.

Indeed, by encouraging the mathematical community to live in relative complacency with respect to foundational issues, the program of set-theoretic foundations may actually be leading us away from fundamental tasks which are clearly of great philosophical importance. The unity of mathematics is valuable, but the unity of human knowledge would be much more valuable.

4. Set-Theoretic Realism

4.1. *An epistemological question*

Some high-level set theorists such as Gödel, Martin, Steel, and Woodin, as well as some high-level philosophers of mathematics such as Maddy, have advocated a philosophical position known as *set-theoretic realism* or Platonism. According to this program, set theory refers to certain definite, undeniable aspects of reality. For instance, cardinals such as $\aleph_\omega$ are thought to exist in a certain domain of reality, and the Continuum Hypothesis is thought to be a meaningful statement about that domain.

An epistemological question which remains is:

How can we acquire knowledge of the set-theoretic reality?

We briefly consider three contemporary answers to this question.

4.2. *The intrinsicist answer*

One answer is that the set-theoretic reality is a non-spatial, non-temporal, irreducible kind of reality which reveals itself by means of pure intuition. I have no response to this, except to say that it seems to express an intrinsicist viewpoint which is obviously incompatible with the requirement of objectivity as I understand it.

4.3. *The "testable consequences" answer*

Another answer to our epistemological question says that the higher set-theoretic reality, although not directly observable, may reveal itself by means of "testable" logical consequences in the concrete mathematical realm. For instance, by Matiyasevich's Theorem, the consistency of a large cardinal axiom can be recast as a number-theoretic statement to the effect that a certain Diophantine equation has no solution in the integers. The resulting justification process for large cardinals is said to be analogous to how the atomic theory of matter was originally discovered and verified, long before it became possible to observe individual atoms directly under an electron microscope.

I find this "testable consequences" viewpoint more appealing than the purely intrinsicist viewpoint, because it gives an active role to a human cognitive process, namely, the study of concrete mathematical problems such as Diophantine equations. Higher set theory is to be justified or reduced or "miniaturized" in terms of its applications to down-to-earth mathematical problems.

The major difficulty that I see with the "testable consequences" program involves its implementation. For instance, the Diophantine equations which have been produced in the manner outlined above are messy and complex and have thousands of terms. No number theorist would seriously study such an equation. Thus, the value of such equations for number theory seems remarkably tenuous. By contrast, the atomic theory from its inception produced a powerful stream of striking consequences in chemistry and other fields of knowledge. These consequences greatly improved the human standard of living.

Attempting to overcome the implementation difficulty, set-theorists have worked very hard for many years trying to uncover consequences of higher set theory and large cardinals which are not only down-to-earth but also *mathematically appealing* and perhaps even *useful in applications*. I am thinking of the impressive results of Martin, Steel and Woodin [2] on projective determinacy,[a] and of Harvey Friedman (unpublished) on Boolean relation theory.

And yet, appealing as they may be, these consequences of large cardinal axioms remain quite remote from standard mathematical practice, especially in application areas. Partly for this reason, they have not led to an upsurge of interest in higher set theory and large cardinals within the mathematical community beyond set theory. Indeed, considering all the hard work that has already gone into this research direction, the prospect of serious impact in core mathematics or in mathematical application areas seems even more unlikely than before.

4.4. *The Thin Realist answer*

Another answer to our epistemological question is Maddy's current philosophy of Thin Realism [3] (in contrast to her earlier Robust Realism, i.e., pure intrinsicism). According to Thin Realism, set theory is in a very strong epistemological position, simply because it is deeply embedded in the "fabric of mathematical fruitfulness." Here again I have my doubts, for the same reasons as above.

Maddy even goes so far as to compare large cardinals to tables and chairs, and set theory skeptics to evil daemon theorists. In other words,

$$\frac{\text{large cardinals}}{\text{set theory skepticism}} = \frac{\text{tables and chairs}}{\text{evil daemon theories}}.$$

---

[a]However, Hugh Woodin notes that this research on projective determinacy was motivated not by the "testable consequences" program, but rather by the desire to answer some long-standing structural questions in the branch of mathematics known as *descriptive set theory* (the study of projective sets in Euclidean space, going back to Souslin and Lusin).

Indeed, according to Maddy, our knowledge of set theory is *more* reliable than our knowledge of tables and chairs, because sense perceptions are subject to skeptical doubts which cannot possibly apply to the "fabric of mathematical fruitfulness."

My view is that, instead of comparing large cardinals to tables and chairs, it seems more appropriate to compare set theory to religion. In other words,

$$\frac{\text{large cardinals}}{\text{set theory skepticism}} = \frac{\text{gods and devils}}{\text{religious skepticism}}.$$

The point of my analogy is that both set theory and religious faith can claim to be in a "strong" position vis a vis skeptics, to the extent that they avoid dependence on underlying facts of reality which can be questioned. In my view, such claims must be rejected on grounds of their lack of objectivity.

Nevertheless, I applaud Maddy's "Second Philosopher" for her earnest attempt to apply standard scientific or epistemological criteria following the lead of other sciences such as biology. It would be very desirable to flesh this out into a full-scale integration of mathematics with the rest of human knowledge.

5. Insights from Reverse Mathematics

For many years I have been involved in a foundational research program known as *reverse mathematics*. The purpose of reverse mathematics is to classify core mathematical theorems according to the set existence axioms which are needed to prove them. Frequently it turns out that a core mathematical theorem is logically equivalent to the weakest such set existence axiom. Hence the name "reverse mathematics." The program has revealed an interesting logical structure within core mathematics. In particular, a large number of core mathematical theorems fall into a small number of logical equivalence classes. Moreover, the set existence axioms which arise in this way are naturally arranged in a hierarchy corresponding roughly to Gödel's hierarchy of consistency strengths. The basic reference on reverse mathematics is my book [6]. Table 1 is from my recent paper [7] and indicates some benchmarks in the Gödel hierarchy.

I believe that many results of reverse mathematics are potentially useful for answering certain questions and evaluating certain programs in the philosophy of mathematics. As regards objectivity in mathematics, I see two insights to be drawn:

Table 1

$$
\text{strong}
\begin{cases}
\vdots \\
\text{supercompact cardinal} \\
\vdots \\
\text{measurable cardinal} \\
\vdots \\
\textsf{ZFC} \text{ (Zermelo/Fraenkel set theory)} \\
\textsf{ZC} \text{ (Zermelo set theory)} \\
\text{simple type theory}
\end{cases}
$$

$$
\text{medium}
\begin{cases}
\textsf{Z}_2 \text{ (second-order arithmetic)} \\
\vdots \\
\Pi_2^1\text{-}\textsf{CA}_0 \ (\Pi_2^1 \text{ comprehension}) \\
\Pi_1^1\text{-}\textsf{CA}_0 \ (\Pi_1^1 \text{ comprehension}) \\
\textsf{ATR}_0 \text{ (arithmetical transfinite recursion)} \\
\textsf{ACA}_0 \text{ (arithmetical comprehension)}
\end{cases}
$$

$$
\text{weak}
\begin{cases}
\textsf{WKL}_0 \text{ (weak König's lemma)} \\
\textsf{RCA}_0 \text{ (recursive comprehension)} \\
\textsf{PRA} \text{ (primitive recursive arithmetic)} \\
\textsf{EFA} \text{ (elementary function arithmetic)} \\
\text{bounded arithmetic} \\
\vdots
\end{cases}
$$

(1) A series of reverse mathematics case studies has shown that the bulk of core mathematical theorems falls at the lowest levels of the hierarchy: $\textsf{WKL}_0$ and below. The full strength of first-order arithmetic appears often but not nearly so often as $\textsf{WKL}_0$. The higher levels up to $\Pi_2^1\text{-}\textsf{CA}_0$ appear sometimes but rarely. For details see [6] and [7].

To me this strongly suggests that higher set theory is, in a sense, largely irrelevant to core mathematical practice. Thus the program of set-theoretic foundations is once again called into question.

(2) It is known that the lowest levels of the Gödel hierarchy (see Table 1) are conservative over PRA (primitive recursive arithmetic) for Π_2^0 sentences. This result combined with reverse mathematics is the basis of some rather strong partial realizations of Hilbert's program of finitistic reductionism, as outlined in my paper [5]. The upshot is that a large portion of core mathematics, sufficient for applications, can be validated by reference to principles which are finitistically provable. It seems to me that these results may open a path toward objectivity in mathematics.

6. Wider Cultural Significance?

Throughout history we see various trends in the philosophy of mathematics, and we see various trends in the culture at large. Are there parallels here? The intrinsicist/subjectivist dichotomy, to which I alluded earlier, may provide some clues.

Clearly mathematics played a large role in the philosophy of Plato and Aristotle and in the Renaissance, the Enlightenment, and the 19th century. However, let us skip ahead to the 20th century.

A thoroughly subjectivistic philosophy of mathematics was Brouwer's Intuitionism. According to Brouwer, mathematics consists of constructions which are performed in the mind of a "creative subject," with no necessary relation to reality. Surely there is a parallel with the subjectivism and collectivism of the early 20th century.

On the intrincist side, consider the rise of religious fundamentalism in the late 20th century: Islamic fundamentalism in the Muslim world, Christian and Jewish fundamentalism in the west, Hindu fundamentalism in India. Could it be that the late 20th century trend toward set-theoretic realism parallels the worldwide rise of religious fundamentalism? This could make an interesting topic of dinner conversation this evening . . . .

Acknowledgment

This paper is the text of my talk at a conference on philosophy of mathematics at New York University, April 3–5, 2009. I wish to thank the NYU Philosophy graduate students and particularly Justin Clarke-Doane and Shieva Kleinschmidt for their attention to detail in organizing the conference. It was exciting to address a wonderful audience at a great urban university in the greatest city in the world.

References

1. S. Feferman, C. Parsons, and S. G. Simpson, editors. *Kurt Gödel: Essays for his Centennial*. Number 33 in Lecture Notes in Logic. Association for Symbolic Logic, Cambridge University Press, 2010. X + 373 pages.

2. Matthew D. Foreman. Review of papers by Donald A. Martin, John R. Steel, and W. Hugh Woodin. *Journal of Symbolic Logic*, 57:1132–1136, 1992.

3. Penelope Maddy. *Second Philosophy: A Naturalistic Method*. Oxford University Press, 2007. XII + 448 pages.

4. Leonard Peikoff. *Objectivism: The Philosophy of Ayn Rand*. Dutton, New York, 1991. XV + 493 pages.

5. Stephen G. Simpson. Partial realizations of Hilbert's program. *Journal of Symbolic Logic*, 53:349–363, 1988.

6. Stephen G. Simpson. *Subsystems of Second Order Arithmetic*. Perspectives in Mathematical Logic. Springer-Verlag, 1999. XIV + 445 pages; Second Edition, Perspectives in Logic, Association for Symbolic Logic, Cambridge University Press, 2009, XVI+ 444 pages.

7. Stephen G. Simpson. The Gödel hierarchy and reverse mathematics. In [1], pages 109–127, 2010.

SORT LOGIC AND FOUNDATIONS OF MATHEMATICS

Jouko Väänänen

Department of Mathematics and Statistics
University of Helsinki, Finland
Institute for Logic, Language and Computation
University of Amsterdam, The Netherlands
jouko.vaananen@helsinki.fi

I have argued elsewhere [8] that second order logic provides a foundation for mathematics much in the same way as set theory does, despite the fact that the former is second order and the latter first order, but second order logic is marred by reliance on ad hoc *large domain assumptions*. In this chapter I argue that sort logic, a powerful extension of second order logic, provides a foundation for mathematics without any ad hoc large domain assumptions. The large domain assumptions are replaced by ZFC-like axioms. Despite this resemblance to set theory sort logic retains the structuralist approach to mathematics characteristic of second order logic. As a model-theoretic logic sort logic is the strongest logic. In fact, every model class definable in set theory is the class of models of a sentence of sort logic. Because of its strength sort logic can be used to formulate particularly strong reflection principles in set theory.

1. Introduction

Sort logic, introduced in [7], is a many-sorted extension of second order logic. In an exact sense it is the strongest logic that there is. In this paper sort logic is suggested as a foundation of mathematics and contrasted to second order logic and to set theory. It is argued that sort logic solves the problem of second order logic that existence proofs of structures rely on ad hoc large domain assumptions.

The new feature in sort logic over and above what first and second order logics have is the ability to "look outside" the model, as for a group to be the multiplicative group of a field requires reference to a zero element outside the group, or for a Turing machine, defined as a finite set of quadruples, to halt requires reference to a tape potentially much bigger than the Turing machine itself.

In computer science it is commonplace to regard a database as a many-sorted structure. Each column (attribute) of the database has its own range of values, be it a salary figure, gender, department, last name, zip code, or whatever. In fact, it would seem very unnatural to lump all these together into one domain which has a mixture of numbers, words, and strings of symbols. To state that a new column can be added to a database, e.g. a salary column, involves stating that new elements, namely the salary values, can be added to the overall set of objects referred to in the database.

In a sense ordinary second order logic also "looks outside" the model as well as one can think of the bound second order variables as first order variables ranging over the domain of all subsets and relations on the original domain. In fact, one of the best ways to understand second order logic is to think of it as a two-sorted first order logic in which one sort—the sort over which the second order variables range—is assumed to consist of *all* subsets and relations of the other sort. When "all subsets and relations" is replaced by "enough subsets and relations to satisfy the Comprehension Axioms", we get semantics relative to which there is a Completeness Theorem of Henkin [2]. The same is true of sort logic.

To get a feeling of sort logic, let us consider the following formulation of the field axioms in a many-sorted first order logic with two sorts of variables. We use variables x, y and z for the sort of the multiplicative group, and u, v and w for the sort of the additive group. The function $\cdot$ and the constant 1 are of the first sort and the function $+$ and the constant 0 of the second sort:

$$
\varphi = \begin{cases}
\forall x \forall y \forall z ((x \cdot y) \cdot z = x \cdot (y \cdot z)) \\
\forall x (x \cdot 1 = 1 \cdot x = x) \\
\forall x \forall y (x \cdot y = y \cdot x) \\
\forall x \exists y (x \cdot y = 1)
\end{cases}
$$

$$
\psi = \begin{cases}
\forall x \forall y \forall z ((x + y) + z = x + (y + z)) \\
\forall x (x + 0 = 0 + x = x) \\
\forall x \forall y (x + y = y + x) \\
\forall x \exists y (x + y = 0) \\
\forall x \forall y \forall z (x \cdot (y + z) = x \cdot y + x \cdot z) \\
\forall x \exists u (x = u) \wedge \forall u \exists x (u = 0 \vee u = x).
\end{cases}
\tag{1.1}
$$

We have separated the multiplicative group into the first sort and the additive group in the second sort. With this separation of the group and the bigger field part we can ask questions such as:

What kind of groups are the multiplicative group of a field?

And the answer is: exactly the groups that satisfy

$$\text{For some } + \text{ and for some } 0: \quad \varphi \wedge \psi. \tag{1.2}$$

The truth of the sentence (1.2) in a given group means that there is something out there outside the group, in this case the element 0, which together with the new function $+$ defines a field.

For a different type of example, suppose

$$\varphi \tag{1.3}$$

is a finite second order axiomatization of some mathematical structure in the vocabulary $\{R_1, ..., R_n\}$. Suppose we want to say that φ has a model. So let us take a new unary predicate P and consider the sentence

$$\exists P(\exists R_1 \ldots \exists R_n \varphi)^{(P)}, \tag{1.4}$$

where $\psi^{(P)}$ means the relativization of ψ to the unary predicate P. What (1.4) says in a model is that there are a subset P and relations $R_1, ..., R_n$ on P such that $(P, R_1, ..., R_n) \models \varphi$. So in any model which is big enough to include a model of φ the sentence (1.4) says that there indeed is such a model. But in smaller models (1.4) is simply false, even though φ may have models. So (1.4) does not really express the existence of a model for φ. The situation would be different if we allowed "$\exists P \exists R_1 \ldots \exists R_n$" to refer to outside the model. In sort logic, which we will introduce in detail below, the meaning of the sentence

$$\tilde{\exists} R_1 \ldots \tilde{\exists} R_n \varphi, \tag{1.5}$$

is that there is a *new* domain of objects with new relations $R_1, ..., R_n$ such that φ holds. Thus (1.5) expresses the semantic consistency of φ independently of the model where it is considered.

In algebra concepts such as a module P being projective, a group F being free, etc, are defined by reference to arbitrary modules M and arbitrary groups G with no concern as to whether such modules M can be realized inside P, or whether such groups G could be realized inside F. Even if it turned out that they could be so realized, the original concepts certainly referred to quite arbitrary objects N and G in the universe of all mathematical objects. Lesson: Apparently second order concepts in mathematics sometimes refer to outside the structure being considered.

Reference outside is, of course, most blatant in set theory where objects are defined by reference to the entire universe of sets. In practice one can in most cases limit the reference to some smaller part of the universe, but very often not to the elements or to the power-set of the object being defined.

2. Sort Logic

Many-sorted logic has several domains, and variables for each domain, much like vector spaces have a scalar-domain and a vector-domain and different variables for each, or as geometry has different variables for points and lines. It seems to have been first considered by Herbrand, and later by Schmidt, Feferman [1], and others.

2.1. *Basic concepts*

A *(many-sorted) vocabulary* is any set L of predicate symbols $P, Q, R, \ldots$. We leave function and constant symbols out for simplicity of presentation. We use natural numbers as names for sorts.

Each vocabulary L has an *arity-function*

$$\mathfrak{a}_L : L \to \mathbb{N}$$

which tells the arity of each predicate symbol, and a *sort-function*

$$\mathfrak{s}_L : L \to \bigcup_n \mathbb{N}^n, \mathfrak{s}_L(R) \in \mathbb{N}^{\mathfrak{a}_L(R)},$$

which tells what are the sorts of the elements of the tuple in a relation. Thus if $P \in L$, then P is an $\mathfrak{a}_L(P)$-ary predicate symbol for a relation of $\mathfrak{a}_L(P)$-tuples of elements of sorts $n_1, \ldots, n_k$, where $(n_1, \ldots, n_k) = \mathfrak{s}(P)$. So we can read off from every n-ary predicate symbol what the sorts of the elements are in the n-tuples of the intended relation. In other words, we do not have symbols for abstract relations between elements of arbitrary sorts (except identity $=$).

2.2. *Syntax*

The syntax of sort logic is very close to the syntax of second order logic. In effect we just add a new form of formula $\tilde{\exists} P \varphi$ with the intuitive meaning that there is a predicate P of *new* sorts of elements such that φ.

Suppose L is a vocabulary. Variable symbols for individuals are $x, y, z, \ldots$ with indexes $x_0, x_1, \ldots$ when necessary, and for relations $X, Y, Z, \ldots$ with indexes $X_0, X_1, \ldots$. Each individual variable x has a sort $\mathfrak{s}(x) \in \mathbb{N}$ associated to it, so it is a variable for elements of sort $\mathfrak{s}(x)$. Each relation variable X has an arity $\mathfrak{a}(X)$ and a sort $\mathfrak{s}(X) \in \mathbb{N}^{\mathfrak{a}(R)}$ associated to it, so it is a relation variable for a relation between elements of the sorts $n_1, \ldots, n_k$, where $\mathfrak{s}(X) = (n_1, \ldots, n_k)$.

The *logical symbols* of sort logic of the vocabulary L are $\approx, \neg, \wedge, \vee, \forall, \exists, (,), x, y, z, \ldots, X, Y, Z, \ldots$. L-*equations* are of the form $x = y$ where x and y can be variables of any sorts. L-*atomic formulas* are either L-equations or of the form $R x_1 \ldots x_k$, where $R \in L$, $\mathfrak{s}_L(R) = (n_1, \ldots, n_k)$, and $x_1, \ldots, x_k$ are individual variables such that $\mathfrak{s}(x_i) = n_i$ for $i = 1, \ldots, k$. A *basic formula* is an atomic formula or the negation of an atomic formula. L-*formulas* are of the form:

(1) $x = y$.
(2) $R(x_1, \ldots, x_n)$, when $\mathfrak{s}_L(R) = (\mathfrak{s}(x_1), \ldots, \mathfrak{s}(x_n))$.
(3) $X(x_1, \ldots, x_n)$, when $\mathfrak{s}(X) = (\mathfrak{s}(x_1), \ldots, \mathfrak{s}(x_n))$.
(4) $\neg\varphi$.
(5) $(\varphi \vee \psi)$.
(6) $\exists x \varphi$.
(7) $\exists X \varphi$.
(8) $\tilde{\exists} X \varphi$. **New Sort Condition**: If $\mathfrak{s}(X) = (n_1, \ldots, n_k)$, then φ has no free variables or symbols of L, other than X, of a sort n_i or of the sort $(m_1, \ldots, m_l)$ with $\{m_1, \ldots, m_l\} \cap \{n_1, \ldots, n_k\} \neq \emptyset$.

The reason for the New Sort Condition is that the domains of the elements referred to by the free variables of φ are fixed already so they should not be altered by the $\tilde{\exists}$-quantifier.

We treat $\varphi \wedge \psi$, $\varphi \to \psi$, $\forall x \varphi$ and $\tilde{\forall} X \varphi$ as shorthands obtained from disjunction and existential quantification by means of negation.

The concept of a free occurrence of a variable in a formula is defined as in first order logic. As a new concept we have the concept of a *free occurrence of a sort* in a formula. We define it as follows, following the intuition that if a sort occurs "free" in a formula, either as the sort of an individual variable, relation variable or predicate symbol, then to understand the meaning of the formula in a model we have to fix the domain of elements of that sort. Respectively, if a sort has only "bound" occurrences in a formula, we can understand the meaning of the formula in a model without fixing the domain of elements of that sort, rather, while evaluating the meaning of the formula in a model we most likely try different domains of elements of that sort.

The *free sorts* $\mathfrak{fs}(\varphi)$ of a formula are defined as follows:

(1) $\mathfrak{fs}(x = y) = \{\mathfrak{s}(x), \mathfrak{s}(y)\}$.
(2) $\mathfrak{fs}(Rx_1 \ldots x_n) = \{\mathfrak{s}(x_1), \ldots, \mathfrak{s}(x_n)\}$.
(3) $\mathfrak{fs}(Xx_1 \ldots x_n) = \{\mathfrak{s}(x_1), \ldots, \mathfrak{s}(x_n)\}$.
(4) $\mathfrak{fs}(\neg\varphi) = \mathfrak{fs}(\varphi)$.
(5) $\mathfrak{fs}(\varphi \vee \psi) = \mathfrak{fs}(\varphi) \cup \mathfrak{fs}(\psi)$.
(6) $\mathfrak{fs}(\exists x\varphi) = \mathfrak{fs}(\varphi) \cup \{\mathfrak{s}(x)\}$.
(7) $\mathfrak{fs}(\exists X\varphi) = \mathfrak{fs}(\varphi) \cup \{n_1, \ldots, n_k\}$, if $\mathfrak{s}(X) = (n_1, \ldots, n_k)$.
(8) $\mathfrak{fs}(\tilde{\exists} X\varphi) = \mathfrak{fs}(\varphi) \setminus \{n_1, \ldots, n_k\}$, if $\mathfrak{s}(X) = (n_1, \ldots, n_k)$.

2.3. *Axioms*

Below $\varphi(y/x)$ means the formula obtained from φ by replacing x by y in its free occurrencies. Substitution should respect sort.

Definition 2.1: The axioms of sort logic are as follows:
Logical axioms:

- Tautologies of propositional logic.
- Identity axioms: $x = y$, $x = y \to y = x$, $(x_1 = y_1 \wedge \ldots \wedge x_n = y_n \wedge \varphi) \to \varphi(y_1...y_n/x_1...x_n)$, for atomic φ
- Quantifier axioms:
    - $\varphi(y/x) \to \exists x\varphi$, if y is free for x in φ in the usual sense.
    - $\varphi(Y/X) \to \exists X\varphi$, if Y is free for X in φ in the usual sense.
    - $\varphi(Y/X) \to \tilde{\exists} X\varphi$, if Y is free for X in φ in the usual sense.

The rules of proof:

- Modus Ponens $\{\varphi, \varphi \to \psi\} \models \psi$
- Generalization
    - $\{\Sigma, \varphi \to \psi\} \models \exists x\varphi \to \psi$, if x is not free in $\Sigma \cup \{\psi\}$
    - $\{\Sigma, \varphi \to \psi\} \models \exists X\varphi \to \psi$, if X is not free in $\Sigma \cup \{\psi\}$
    - $\{\Sigma, \varphi \to \psi\} \models \tilde{\exists} X\varphi \to \psi$, if no free sorts of ψ occur in $\mathfrak{s}(X)$.

First Comprehension Axiom:

$$\exists X\forall y_1...\forall y_m(Xy_1...y_m \leftrightarrow \psi)$$

for any formula ψ not containing X free, whenever $\mathfrak{s}(X) = (\mathfrak{s}(y_1), \ldots, \mathfrak{s}(y_m))$.

Second Comprehension Axiom:

$$\tilde{\exists}X\forall y_1...\forall y_m(Xy_1...y_m \leftrightarrow \psi)$$

for any formula ψ not containing X free, whenever $\mathfrak{s}(X) = (\mathfrak{s}(y_1),\ldots,\mathfrak{s}(y_m))$.

The logical axioms and the rules of proof are clearly indispensable and are directly derived from corresponding axioms and rules of first order logic. The difference between the axioms $\varphi(Y/X) \to \exists X\varphi$ and $\varphi(Y/X) \to \tilde{\exists}X\varphi$ is the following: Both take $\varphi(Y/X)$ as a hypothesis. The conclusion $\exists X\varphi$ says of the current sorts that a relation X satisfying φ exists, namely, Y. If $\mathfrak{s}(X) = (n_1,\ldots,n_k)$, then the conclusion $\tilde{\exists}X\varphi$ says of the sorts other than $n_1,\ldots,n_k$ that domains for the sorts $n_1,\ldots,n_k$ exists so that in the combined structure of the old and new domains a relation X satisfying φ exists, namely, Y. The Comprehension Axiom is the traditional (impredicative) axiom schema which gives second order logic, and in our case sort logic, the necessary power to do mathematics [3]. In individual cases less comprehension may be sufficient but this is the general schema. The difference between the First and the Second Comprehension Axioms is that the former stipulates the existence of a relation X defined by ψ in the structure consisting of the existing sorts, while the latter says that this is even true if the sorts of elements and relations that ψ

If we limit ourselves to just one sort, for example 0, we get exactly the classical second order logic.

2.4. *Semantics*

We now define the semantics of sort logic. This is very much like the semantics of second order logic, except that we have to take care of the new domains that may arise from interpreting quantifiers of the form $\tilde{\exists}$ and $\tilde{\forall}$.

Definition 2.2: An *L-structure* (or *L-model*) is a function $\mathcal{M}$ defined on L with the following properties:

(1) If $R \in L$ and $\mathfrak{s}(R) = (n_1,...,n_k)$ then $n_i \in \text{dom}(\mathcal{M})$ and $M_{n_i} =_{df} \mathcal{M}(n_i)$ is a non-empty set for each $i \in \{1,...,k\}$.
(2) If $R \in L$ is an k-relation symbol and $\mathfrak{s}(R) = (n_1,...,n_k)$, then $\mathcal{M}(R) \subseteq M_{n_1} \times \ldots \times M_{n_k}$.

We usually shorten $\mathcal{M}(R)$ to $R^{\mathcal{M}}$. If no confusion arises, we use the notation

$$\mathcal{M} = (M_{n_1},\ldots,M_{n_l}; R_1^{\mathcal{M}},\ldots,R_m^{\mathcal{M}})$$

for a many-sorted structure with universes $M_{n_1}, \ldots, M_{n_l}$ and relations $R_1^{\mathcal{M}}, \ldots, R_m^{\mathcal{M}}$ between elements of some of the universes. A vector space with scalar field F and vector group V would be denoted according to this convention (taking functions and constant relationally):

$$(V, F; \; \cdot \, , 1, \; + \, , 0).$$

Definition 2.3: An *assignment* into an L-structure $\mathcal{M}$ is any function s the domain of which is a set of individual variables, relation variables and natural numbers such that

(1) If $x \in \text{dom}(s)$, then $s(x) \in \text{dom}(\mathcal{M})$ and $s(x) \in M_{s(x)}$.
(2) If $X \in \text{dom}(s)$ with $s_L(X) = (n_1, \ldots, n_k)$, then $n_1, \ldots, n_k \in \text{dom}(\mathcal{M})$ and $s(X) \subseteq M_{n_1} \times \ldots \times M_{n_k}$.

A *modified assignment* is defined as follows:

$$s[a/x](y) = \begin{cases} a & \text{if } y = x \\ s(y) & \text{otherwise.} \end{cases}$$

$$s[A/X](Y) = \begin{cases} A & \text{if } Y = X \\ s(Y) & \text{otherwise.} \end{cases}$$

Suppose $s(X) = (n_1, \ldots, n_k)$. A model $\mathcal{M}'$ is an X-*expansion* of a model $\mathcal{M}$ if $\{n_1, \ldots, n_k\} \cap \text{dom}(\mathcal{M}) = \emptyset$, $\text{dom}(\mathcal{M}') = \text{dom}(\mathcal{M}) \cup \{n_1, \ldots, n_k\}$, and $\mathcal{M}' \upharpoonright \text{dom}(\mathcal{M}) = \mathcal{M}$.

Definition 2.4: The *truth* of L-formulas in $\mathcal{M}$ under s is defined as follows:

(1) $\mathcal{M} \models_s R(x_1, \ldots, x_n)$ if and only if $(s(x_1), \ldots, s(x_n)) \in \mathcal{M}(R)$,
(2) $\mathcal{M} \models_s x = y$ if and only if $s(x) = s(y)$,
(3) $\mathcal{M} \models_s \neg\varphi$ if and only if $\mathcal{M} \nvDash_s \varphi$,
(4) $\mathcal{M} \models_s (\varphi \vee \psi)$ if and only if $\mathcal{M} \models_s \varphi$ or $\mathcal{M} \models_s \psi$,
(5) $\mathcal{M} \models_s \exists x\varphi$ if and only if $\mathcal{M} \models_{s[a/x]} \varphi$ for some $a \in M_{s(x)}$,
(6) $\mathcal{M} \models_s \exists X\varphi$ if and only if $\mathcal{M} \models_{s[A/X]} \varphi$ for some $A \subseteq M_{n_1} \times \ldots \times M_{n_k}$, where $s(X) = (n_1, \ldots, n_k)$,
(7) $\mathcal{M} \models_s \tilde{\exists} X\varphi$ if and only if $\mathcal{M}' \models_{s[A/X]} \varphi$ for some X-expansion $\mathcal{M}'$ of $\mathcal{M}$ and some $A \subseteq M'_{n_1} \times \ldots \times M'_{n_k}$, where $s(X) = (n_1, \ldots, n_k)$.

Since (7) of the above truth definition involves unbounded quantifiers over sets, the definition has to be given separately for formulas of quantifier-rank at most a fixed natural number n. When n increases, the definition itself gets more complicated in the sense of the quantifier rank.

As in second order logic, there is a looser concept of a model, one relative to which we can prove a Completeness Theorem. This concept permits also a uniform definition.

Definition 2.5: A *Henkin L-structure* (or Henkin *L-model*) is a triple $(\mathcal{M},\mathcal{U},\mathcal{G})$, where $\mathcal{M}$ is an L-structure, $\mathcal{U}$ is a set such that $\emptyset \notin \mathcal{U}$ and $\mathcal{G}$ is a set of relations between elements of the domains of $\mathcal{M}$ and the sets in $\mathcal{U}$. We assume that the First and the Second Comprehension Axioms are satisfied by $(\mathcal{M},\mathcal{U},\mathcal{G})$ in the sense defined below.

The idea is that $\mathcal{U}$ gives a set of possible domains for the new sorts needed for the truth conditions of the $\tilde{\exists}$-quantifiers, and $\mathcal{G}$ gives a set of possible relations needed for the truth conditions of the $\exists$-quantifiers. Since $\mathcal{U}$ is not the class of all sets (as it is a set) and $\mathcal{G}$ need not be the set of *all* relevant relations, the structures $(\mathcal{M},\mathcal{U},\mathcal{G})$ are more general than the structures $\mathcal{M}$. The original structures $\mathcal{M}$ are called *full*.

An *assignment* and a *modified assignment* for a Henkin L-structure $(\mathcal{M},\mathcal{U},\mathcal{G})$ is defined as for ordinary structures. Suppose $\mathfrak{s}(X) = (n_1,\ldots,n_k)$. A model $\mathcal{M}'$ is an *X-expansion in $\mathcal{U}$* of a model $\mathcal{M}$ if $\{n_1,\ldots,n_k\} \cap \mathrm{dom}(\mathcal{M}) = \emptyset$, $\mathrm{dom}(\mathcal{M}') = \mathrm{dom}(\mathcal{M}) \cup \{n_1,\ldots,n_k\}$, $\mathcal{M}' \upharpoonright \mathrm{dom}(\mathcal{M}) = \mathcal{M}$, and $\mathcal{M}'(n_i) \in \mathcal{U}$ for all $i = 1,\ldots,k$.

Definition 2.6: The *truth* of L-formulas in $(\mathcal{M},\mathcal{U},\mathcal{G})$ under s is defined as follows:

(1) $(\mathcal{M},\mathcal{U},\mathcal{G}) \models_s R(x_1,\ldots,x_n)$ if and only if $(s(x_1),\ldots,s(x_n)) \in \mathcal{M}(R)$,

(2) $(\mathcal{M},\mathcal{U},\mathcal{G}) \models_s x = y$ if and only if $s(x) = s(y)$,

(3) $(\mathcal{M},\mathcal{U},\mathcal{G}) \models_s \neg\varphi$ if and only if $(\mathcal{M},\mathcal{U},\mathcal{G}) \not\models_s \varphi$,

(4) $(\mathcal{M},\mathcal{U},\mathcal{G}) \models_s (\varphi \vee \psi)$ if and only if $(\mathcal{M},\mathcal{U},\mathcal{G}) \models_s \varphi$ or $(\mathcal{M},\mathcal{U},\mathcal{G}) \models_s \psi$,

(5) $(\mathcal{M},\mathcal{U},\mathcal{G}) \models_s \exists x\varphi$ if and only if $(\mathcal{M},\mathcal{U},\mathcal{G}) \models_{s[a/x]} \varphi$ for some $a \in M_{\mathfrak{s}(x)}$,

(6) $(\mathcal{M},\mathcal{U},\mathcal{G}) \models_s \exists X\varphi$ if and only if $(\mathcal{M},\mathcal{U},\mathcal{G}) \models_{s[A/X]} \varphi$ for some $A \in \mathcal{P}(M_{n_1} \times \ldots \times M_{n_k}) \cap \mathcal{G}$, where $\mathfrak{s}(X) = (n_1,\ldots,n_k)$,

(7) $(\mathcal{M},\mathcal{U},\mathcal{G}) \models_s \tilde{\exists} X\varphi$ if and only if $(\mathcal{M},\mathcal{U},\mathcal{G})' \models_{s[A/X]} \varphi$ for some X-expansion $\mathcal{M}'$ of $\mathcal{M}$ in $\mathcal{U}$ and some $A \in \mathcal{P}(M'_{n_1} \times \ldots \times M'_{n_k}) \cap \mathcal{G}$, where $\mathfrak{s}(X) = (n_1,\ldots,n_k)$.

The following characterization of provability in sort logic is proved as the corresponding result for type theory [2]:

Theorem 2.7: *(Completeness Theorem) The following conditions are equivalent for any sentence φ of sort logic and any countable theory T of sort logic:*

(1) $T \models \varphi$.
(2) Every Henkin model of T satisfies φ.
(3) Every countable Henkin model of T satisfies φ.

This characterization shows that our axioms for sort logic capture the intuition of sort logic in a perfect manner, at least if our Henkin semantics does. Our Henkin semantics is very much like that of second order logic.

3. Sort Logic and Set Theory

In this chapter we look at sort logic from the point of vies of set theory.

Definition 3.1: We use Δ_n to denote the set of formulas of sort logic which are (semantically) equivalent both to a Σ_n-formula of sort logic, and to a Π_n-formula of sort logic.

Theorem 3.2: [7][6] *The following conditions are equivalent for any model class K and for any $n > 1$:*

(1) K is definable in the logic Δ_n.
(2) K is Δ_n-definable in the Levy-hierarchy.

Proof: We give the proof only in the case $n = 2$. The general case is similar. Suppose L is a finite vocabulary and $\mathcal{A}$ is a second order characterizable L-structure. Suppose σ is the conjunction of a large finite part of ZFC. Let us call a model $(M, \in)$ of θ *supertransitive* if for every $a \in M$ every element and every subset of a is in M. Let $\mathrm{Sut}(M)$ be a Π_1-formula which says that M is supertransitive. Let $\mathrm{Voc}(x)$ be the standard definition of "x is a vocabulary". Let $\mathrm{SO}(L, x)$ be the set-theoretical definition of the class of second order L-formulas. Let $\mathrm{Str}(L, x)$ be the set-theoretical definition of L-structures. Let $\mathrm{Sat}(\mathcal{A}, \varphi)$ be an inductive truth-definition of the Σ_2-fragment of sort logic written in the language of set theory. Let

$$P(z, x, y) = \mathrm{Voc}(z) \wedge \mathrm{Str}(z, x) \wedge \mathrm{SO}(z, y) \wedge$$
$$\exists M (z, x, y \in M \wedge \sigma^{(M)} \wedge \mathrm{Sut}(M) \wedge (\mathrm{Sat}(z, x, y))^{(M)}).$$

Now if L is a vocabulary, $\mathcal{A}$ an L-structure, then $\mathcal{A} \models \varphi \iff P(L, \mathcal{A}, \varphi)$. This shows that $\mathcal{A} \models \varphi$ is a Σ_2 property of $\mathcal{A}$ and L.

For the converse, suppose the predicate Φ is a Σ_2 property of L-structures. There is a Σ_2-sentence φ of sort logic such that for all $\mathcal{M}$, $\mathcal{M} \in K$ off $\mathcal{M} \models \varphi$.

Suppose $\Phi = \exists x \forall y P(x, y, \mathcal{A})$, a Σ_2-property of $\mathcal{A}$. Let ψ be a sort logic sentence the models of which are, up to isomorphism, exactly the models $\mathcal{A}$ for which there is $(V_\alpha, \in)$, with $\alpha = \beth_\alpha$, $\mathcal{A} \in V_\alpha$, and $(V_\alpha, \in) \models \exists x \forall y P(x, y, \mathcal{A})$. If $\exists x \forall y P(x, y, \mathcal{A})$ holds, we can find a model for ψ by means of the Levy Reflection principle. On the other hand, suppose ψ has a model $\mathcal{A}$. W.l.o.g. it is of the form $(V_\alpha, \in)$ with $\mathcal{A} \in V_\alpha$. Let $a \in V_\alpha$ such that $(V_\alpha, \in) \models \forall y P(a, y, \mathcal{A})$. Since in this case $H_\alpha = V_\alpha$, $(H_\alpha, \in) \models \forall y P(a, y, \mathcal{A})$, where H_α is the set of sets of hereditary cardinality $< \alpha$. By another application of the Levy Reflection Principle we get $(V, \in) \models \forall y P(a, y, \mathcal{A})$, and we have proved $\exists x \forall y P(x, y, \mathcal{A})$. $\qquad \square$

By a *model class* we mean a class of structures of the same vocabulary, which is closed under isomorphisms. In the context of set theory classes are referred to by their set-theoretical definitions.

The following consequence was mentioned in [5] without proof:

Corollary 3.3: [7][6] *Every model class is definable in sort logic. Sort logic is therefore the strongest logic.*

The logics Δ_n, $n = 2, 3, \ldots$, provide a sequence of stronger and stronger logics. Their model theoretic properties can be characterized in set theoretical terms as the following results indicate:

Theorem 3.4: [7] *The Hanf-number of the logic Δ_n is δ_n. The Löwenheim number of the logic Δ_n is σ_n. The decision problem of the logic Δ_n is the complete Π_n-definable set of natural numbers.*

Theorem 3.5: [4] *The LST-number of Δ_2 is the first supercompact cardinal.*

Theorem 3.6: *The LST-number of Δ_3 is at least the first extendible cardinal.*

Theorem 3.7: *The decision problem of Δ_n is the complete Π_n-set of natural numbers.*

4. Sort Logic and Foundations of Mathematics

We suggest that sort logic can provide a foundation of mathematics in the same way as second order logic, with the strong improvement that it does not depend on the ad hoc Large Domain Assumptions of set theory.

We will now think of foundations of mathematics from the point of view of sort logic. Propositions of mathematics are—according to the sort logic view—either of the form

$$\mathcal{A} \models \varphi,$$

where $\mathcal{A}$ is a structure, characterizable in sort logic, and φ is a sentence of sort logic, or else of the form

$$\models \varphi,$$

where again φ is a sentence of sort logic. Thus a proposition of mathematics either states a *specific* truth, truth in a specific structure, or a *general* truth, truth in all structures. The specific truth can be reduced to the general truth as follows. Suppose $\theta_{\mathcal{A}}$ is a sort logic sentence which characterizes $\mathcal{A}$ up to isomorphism. Then

$$\mathcal{A} \models \varphi \iff \models \theta_{\mathcal{A}} \to \varphi.$$

Curiously, and quite unlike the case of second order logic, the converse holds, too. Suppose φ is a sort logic sentence in which the predicates $P_1, ..., P_k$ occur only. Let $X_1, \ldots, X_k$ be new unary predicate variables of sorts $\mathfrak{s}(P_1), ..., \mathfrak{s}(P_k)$ respectively. Then

$$\models \varphi \iff \mathcal{A} \models \tilde{\forall} X_1 \ldots \tilde{\forall} X_k \varphi(X_1 \ldots X_k / P_1 \ldots P_k)$$

$$\not\models \varphi \iff \mathcal{A} \models \neg \tilde{\forall} X_1 \ldots \tilde{\forall} X_k \varphi(X_1 \ldots X_k / P_1 \ldots P_k).$$

Intuitively this says that a sort logic sentence which talks about the predicates $P_1, \ldots, P_k$ in some domains is valid if and only if whatever new domains and interpretations we take for $P_1, \ldots, P_k$, φ is true. Since the general truth is reducible to specific truth we may focus on specific truth only, without loss of generality.

What is the justification we can give to asserting $\mathcal{A} \models \varphi$? We can *prove* from the axioms of sort logic the sentence $\theta_{\mathcal{A}} \to \varphi$. Of course, we may have to go beyond the standard axioms of sort logic, but much of mathematics can be justified with the sort logic axioms that we have.

What is the justification for asserting that φ has a model, i.e. $\neg \varphi$ is not valid? By the above it suffices to prove from the axioms the sentence $\neg \tilde{\forall} X_1 \ldots \tilde{\forall} X_k \varphi(X_1 \ldots X_k / P_1 \ldots P_k)$. If we compare the situation of sort logic with that of second order logic the difference is that in second order logic we have to make so-called "large domain assumptions" to justify existence of mathematical structures, while in sort logic we can simply prove them from the general Comprehension Axioms. But here comes a moment of truth. Can we actually prove the existence of the structures necessary in mathematics from the mere Comprehension Axioms?

In fact we need more axioms to supplement the First and the Second Comprehension Axiom.

Definition 4.1: Power Sort Axiom:

$$\tilde{\exists}Y(\forall u\exists z_1(u = z_1) \wedge \forall z_1\exists u(u = z_1)\wedge$$
$$\forall x\forall y(\forall z_1\ldots\forall z_n(Yxz_1\ldots z_n \leftrightarrow Yyz_1\ldots z_n) \to x = y)\wedge \qquad (4.1)$$
$$\forall X\exists x\forall z_1\ldots\forall z_n(Xz_1\ldots z_n \leftrightarrow Yxz_1\ldots z_n)),$$

where

$$\mathfrak{s}(X) = (\mathfrak{s}(z_1),\ldots,\mathfrak{s}(z_n))$$
$$\mathfrak{s}(Y) = (\mathfrak{s}(x),\mathfrak{s}(z_1),\ldots,\mathfrak{s}(z_n))$$
$$\mathfrak{s}(x) = \mathfrak{s}(y)$$
$$\mathfrak{s}(z_1) = \ldots = \mathfrak{s}(z_n).$$

Note that in the Power Sort Axiom only $\mathfrak{s}(u)$ occurs free, so it is an axiom about models with one sort, namely $\mathfrak{s}(u)$. Naturally the models may consist of other sorts as well, this axiom just does not say anything about those sorts. The sort $\mathfrak{s}(z_1)$ is just an auxiliary copy of the sort $\mathfrak{s}(u)$, as the conjunct $\forall u\exists z_1(u = z_1) \wedge \forall z_1\exists u(u = z_1)$ stipulates. The sort $\mathfrak{s}(x)$ is a new sort which codes the n-ary relations on the domain $\mathfrak{s}(u)$. The coding is done by means of the predicate Y.

Lemma 4.2: *Every full model satisfies the Power Sort Axiom.*

Proof: Suppose $\mathcal{M}$ is a (full) model and s is an assignment into $\mathcal{M}$. Let us fix X, Y, and $x, y, z_1,\ldots,z_n$ such that $\mathfrak{s}(X) = (\mathfrak{s}(z_1),\ldots,\mathfrak{s}(z_n))$, $\mathfrak{s}(Y) = (\mathfrak{s}(x),\mathfrak{s}(z_1),\ldots,\mathfrak{s}(z_n))$, $\mathfrak{s}(x) = \mathfrak{s}(y)$ and $\mathfrak{s}(z_1) = \ldots = \mathfrak{s}(z_n)$. Let $\mathcal{N}$ be like $\mathcal{M}$ except that there is a new sort $\mathfrak{s}(u)$ (or if this sort existed in $\mathcal{M}$ it is now replaced) with universe $\mathcal{P}(M_{\mathfrak{s}(x)})$ and $\mathcal{N}_{\mathfrak{s}(z_1)} = M_{\mathfrak{s}(u)}$. Let s' be like s except that

$$s'(Y) = \{(a,b_1,\ldots,b_n) \in N_{\mathfrak{s}(x)} \times M_{\mathfrak{s}(u)} \times \ldots \times M_{\mathfrak{s}(u)} : (b_1,\ldots,b_n) \in a\}.$$

Now s' satisfies in $\mathcal{N}$ the formula

$$\forall x\forall y(\forall z_1\ldots\forall z_n(Yxz_1\ldots z_n \leftrightarrow Yyz_1\ldots z_n)) \to x = y)\wedge$$
$$\forall X\exists x\forall z_1\ldots\forall z_n(Xz_1\ldots z_n \leftrightarrow Yxz_1\ldots z_n)),$$

Hence s satisfies in $\mathcal{M}$ the formula (4.1). $\qquad\square$

Definition 4.3: Infinite Sort Axiom:

$$\tilde{\exists}X(\forall x\forall y\forall z((Xxy \wedge Xxz) \rightarrow y = z)$$
$$\forall x\forall y\forall z((Xxz \wedge Xyz) \rightarrow x = y) \qquad (4.2)$$
$$\forall x\exists y Xxy$$
$$\exists z\forall x\forall y(Xxy \rightarrow \neg y = z))$$

where $\mathfrak{s}(X) = (\mathfrak{s}(x), \mathfrak{s}(x))$ and $\mathfrak{s}(x) = \mathfrak{s}(y) = \mathfrak{s}(z)$.

Lemma 4.4: *Every full model satisfies the Infinite Sort Axiom.*

Proof: Suppose $\mathcal{M}$ is a (full) model and s is an assignment into $\mathcal{M}$. Let us fix X, and x, y, z such that $\mathfrak{s}(X) = (\mathfrak{s}(x), \mathfrak{s}(x))$, and $\mathfrak{s}(x) = \mathfrak{s}(y) = \mathfrak{s}(z)$. Let $\mathcal{M}'$ be like $\mathcal{M}$ except that there is a new sort $\mathfrak{s}(x)$ (or if this sort existed in $\mathcal{M}$ it is now replaced) with universe ω. Let s' be like s except that

$$s'(X) = \{(n, n+1) : n \in \omega\}.$$

Now s' satisfies in $\mathcal{M}'$ the formula

$$\forall x\forall y\forall z((Xxy \wedge Xxz) \rightarrow y = z)$$
$$\forall x\forall y\forall z((Xxz \wedge Xyz) \rightarrow x = y)$$
$$\forall x\exists y Xxy$$
$$\exists z\forall x\forall y(Xxy \rightarrow \neg y = z).$$

Hence s satisfies in $\mathcal{M}$ the formula (4.2). $\qquad \square$

The Power Sort Axiom, reminiscent of the Power Set Axiom of set theory, is necessary for arguing about the existence of new sorts of elements. The Infinite Sort Axiom is required for arguing about infinite domains, just as we need the Axiom of Infinity in set theory. Note that the Power Sort Axiom or the Infinite Sort Axiom do not imply that we have only infinite or uncountable models. By the above lemmas these axioms are true in all models, even finite ones.

With the above two new axioms we can construct mathematical structures up to any cardinality $< \beth_\omega$, as if we were working in Zermelo's set theory. For bigger structures we have to make stronger assumptions, and they probably have great similarity with the Replacement Axiom of set theory. The point is that in second (and higher) order logic we have to make ad hoc Large Domain Assumptions as we go from structure to structure, while in sort logic we need only make general assumptions about domains, as axioms of set theory postulate general properties of sets.

So what is the difference between sort logic and set theory? Despite its proximity to set theory, sort logic is still a logic, like first order logic, second order logic, infinitary logic, etc. Sort logic treats mathematical structures up to isomorphism only, there is no preference of one construction of a structure over another, and this is in line with common mathematical thinking about structures. In its model theoretic formulation sort logic gives rise to interesting reflection principles via its Löwenheim-Skolem-Tarski properties. Finally, sort logic provides a natural model theoretic forum for investigating complicated set theoretical properties of models, without going into the nuts and bolts of the constructions of specific structures.

As the strongest logic sort logic is an ultimate yard-stick of definability in mathematics. Any property that is isomorphism invariant can be measured by sort logic. The canonical hierarchy Δ_n ($n < \omega$) inside sort logic climbs up the large cardinal hierarchy by reference to Hanf-, Löwenheim- and Skolem-Löwenhem-Tarski-numbers, reaching all the way to Vopenka's Principle. The model classes that are Δ_2 in the Levy-hierarchy are exactly the model classes definable in the Δ-extension of second order logic. Sort logic provides a similar characterization of model classes that are Δ_n in the Levy-hierarchy for $n > 2$. Is this too strong a logic to be useful? For logics as strong as sort logic the main use is in definability theory. But sort logic has also a natural axiomatization, complete with respect to a natural concept of a Henkin model, so we can also write inferences in sort logic. This is an alternative way of looking at mathematics to set theory, one in which definition rather than construction is the focus.

References

1. Solomon Feferman. Applications of many-sorted interpolation theorems. In *Proceedings of the Tarski Symposium (Proc. Sympos. Pure Math., Vol. XXV, Univ. of California, Berkeley, Calif., 1971)*, pages 205–223, Providence, R. I., 1974. Amer. Math. Soc.
2. Leon Henkin. Completeness in the theory of types. *J. Symbolic Logic*, 15:81–91, 1950.
3. David Hilbert and Wilhelm Ackermann. *Grundzüge der theoretischen Logik*. Springer-Verlag, Berlin, 1972. Sixth edition, Die Grundlehren der mathematischen Wissenschaften, Band 27. First edition published 1928.
4. M. Magidor. On the role of supercompact and extendible cardinals in logic. *Israel J. Math.*, 10:147–157, 1971.
5. J. A. Makowsky, Saharon Shelah, and Jonathan Stavi. Δ-logics and generalized quantifiers. *Ann. Math. Logic*, 10(2):155–192, 1976.
6. Juha Oikkonen. Second-order de.nability, game quantifiers and related expressions. *Soc. Sci. Fenn. Comment. Phys.-Math.*, 48(1):39–101, 1978.
7. Jouko Väänänen. Abstract logic and set theory. I. Definability. In *Logic Colloquium 78 (Mons, 1978)*, volume 97 of *Stud. Logic Foundations Math.*, pages

391–421. North-Holland, Amsterdam, 1979.

8. Jouko Väänänen. Second-order logic or set theory? *Bull. Symbolic Logic*, 18(1):91–121, 2012.

REASONING ABOUT CONSTRUCTIVE CONCEPTS

Nik Weaver

Department of Mathematics
Washington University in St. Louis
St. Louis, MO 63130, USA
nweaver@math.wustl.edu

We find that second order quantification is problematic when a quantified concept variable is supposed to function predicatively. This issue is analyzed and it is shown that a constructive interpretation of the falling under relation suffices to resolve the difficulty. We are then able to present a formal system for reasoning about concepts. We prove that this system is consistent and we investigate the extent to which it is able to interpret set theoretic and number theoretic systems of a more standard type.

1.

Second order quantification becomes problematic when a quantified concept variable is supposed to function predicatively. There is a use/mention issue.

The distinction that comes into play is illustrated in Tarski's classic biconditional [2]

"Snow is white" is true $\leftrightarrow$ snow is white.

The expression "snow is white" is mentioned on the left side; there it is linguistically inert and appears only as an object under discussion. On the right side it is in use and has assertoric force.

To see the sort of problem that can arise, suppose we try to define the truth of an arbitrary sentence by saying

$$\ulcorner A \urcorner \text{ is true} \leftrightarrow A, \tag{$*$}$$

where A is taken to range over all sentences and the use/mention distinction

187

is indicated using corner brackets. Why is this one statement not a global definition of truth?

The answer depends on whether the variable A in $(*)$ is understood as schematic or as being implicitly quantified. If we interpret $(*)$ schematically, that is, as a sort of template which is not itself an assertion but which becomes one when any sentence is substituted for A, then it cannot be a definition of truth since it is the wrong kind of object (a definition can be asserted, a template cannot). We could still use it as a tool to construct truth definitions in limited settings: given any target language, the conjunction of all substitution instances of $(*)$, as A ranges over all the sentences of the language, would define truth for that language. But this conjunction generally will not belong to the target language, so we cannot construct a global definition of truth for all sentences in this way. More to the point, we cannot use this approach to build a language in which we have the ability to discuss the truth of any sentence in that language.

The expression $(*)$ could be directly interpreted as a truth definition for all sentences only by universally quantifying the variable A. However, this is impossible for straightforward syntactic reasons. In the quantifying phrase "for every sentence A" the symbol A has to represent a mention, not a use, of an arbitrary sentence, since here the arbitrary sentence is being referred to and not asserted. But we need A to represent a use in the right side of the biconditional. So there is no meaningful way to quantify over A in $(*)$. This expression can only be understood as schematic.

(The general principle is that a schematic expression can be obtained by omitting any part of a well-formed sentence, but we can only quantify over omitted noun phrases. "Snow is white" can be schematized to either "x is white" or "Snow C", but "$(\exists x)(x$ is white$)$" is grammatical while "$(\exists C)(\text{snow } C)$" is not. In $(*)$ the omission is not nominal.)

Since a quantified variable can only represent a mention of an arbitrary sentence, what we would need in order to formulate a global definition of truth is a disquotation operator $\ulcorner \cdot \urcorner$. Then we could let the variable A refer to an arbitrary sentence and write

For every sentence A, A is true $\leftrightarrow \ulcorner A \urcorner$.

In other words, we need some way to convert mention into use. But that is exactly what having a truth predicate does for us. The way we convert a mention of the sentence "snow is white" into an actual assertion that snow is white is by saying that the mentioned sentence is true. So in order to state a global definition of truth we would need to, in effect, already have a global notion of truth.

This is why no variant of (∗) can succeed in globally defining truth. A quantified variable representing an arbitrary sentence in general cannot be invested with assertoric force unless we possess a notion of truth that applies to all sentences, but writing down a global definition of truth requires us to already be able to construe such a variable as having assertoric force.

Truth seems unproblematic because in any particular instance it really is unproblematic. Any meaningful sentence can be substituted for A in (∗) with straightforward results. But the essentially grammatical problem with quantifying over A in (∗) is definitive. There is no way to use this schematic condition to globally define truth, and we can be quite certain of this because any predicate which globally verified (∗) would engender a contradiction. It would give rise to a liar paradox.

2.

Similar comments can be made about what it means for an object to fall under a concept. Just as with truth, there is no difficulty in defining this relation in any particular case. For instance, we can define what it means to fall under the concept *white* by saying

> The object x falls under the concept *white* $\leftrightarrow$ x is white.

But if we try to characterize the falling under relation globally by saying

$$\text{The object } x \text{ falls under the concept } \ulcorner C \urcorner \leftrightarrow Cx \qquad (\dagger)$$

then, just as with (∗), a use/mention conflict arises when we try to quantify over C. We can use (†) as a template to produce a falling under definition for any particular concept; we can even take the conjunction of the substitution instances of (†) as C ranges over all the concepts expressible in a given language, and thereby obtain a falling under definition for that language, but this definition could not itself belong to the language in question. As with (∗), in order to put (†) in a form that would allow C to be quantified, so that it could have global force, we would need some device for converting a mention of an arbitrary concept into a use of that concept. But that is exactly what the falling under relation does for us. That is to say, we need to already have a global notion of falling under before we can use (†) to define falling under globally. Thus, no variant of (†) can succeed in globally defining a falling under relation.

In fact, the twin difficulties with truth and falling under are not just analogous, they are effectively equivalent. If we had a globally applicable truth predicate, we could use it to define a global notion of falling under, viz., "An object falls under the concept $C \leftrightarrow$ the atomic proposition formed from a name for that object and C is true." Conversely, given a globally applicable falling under relation, truth could be defined globally by saying "The sentence A is true $\leftrightarrow$ every object falls under the predicate formed by concatenating 'is such that' with A". We can now see that truth and falling under are practically identical notions. Falling under is to formulas with one free variable what truth is to sentences.

However, there is one striking difference between the two cases. The globally problematic nature of truth does not have any immediate implications for our understanding of logic, but, in contrast, the globally problematic nature of falling under has severe consequences for general second order quantification. As we have seen, expressions like $(\exists C)Cx$ are, taken at face value, syntactically ill-formed. The quantified concept variable C cannot function predicatively because its appearance in the quantifying phrase is a mention, not a use. In order to make sense of expressions like this we need a global device for converting a mention of an arbitrary concept into a use of that concept, which is just to say that we need a global notion of falling under. But it should now be clear that a global notion of falling under, in the form of a relation which satisfies (†) for every concept C, is something we do not and cannot have. (Cannot, because it would give rise to Russell's paradox.) Now, if C were restricted to range over only those concepts appearing in some given language, then we could use (†) as a template to define falling under for those concepts and thereby render quantification over them meaningful. But sentences employing these quantifiers would not belong to the target language, so this approach cannot be used to make sense of unrestricted second order quantification. Specifically, it cannot be used to build a language in which we have the ability to quantify over all concepts expressible in that language while allowing the quantified concept variable to predicate.

Thus, we are not straightforwardly able to assign meaning to statements in which a quantified concept variable is supposed to function predicatively.

3.

This negative conclusion is unsatisfying because our syntactic considerations forbid not only the paradoxical global definitions of truth and falling under which we want to exclude, but also other global statements which appear to be meaningful. For instance, we have remarked that in limited settings truth and falling under are unproblematic, but it is not obvious how to formalize this claim itself. We cannot say that for any sentence A there is a predicate T such that $T(A) \leftrightarrow A$, for the same reason that we cannot quantify over A in (∗): the mentions of T and A in the quantifying phrases are followed by uses in the expression $T(A) \leftrightarrow A$. Reformulations like "such that $T(A) \leftrightarrow A$ holds" or "such that $T(A) \leftrightarrow A$ is the case" accomplish nothing because they merely employ synonyms for truth. But this prohibition is confusing because there clearly is some sense in which it is correct, even trivially correct, to say that a truth definition can be given for any meaningful sentence. We either have to adopt the mystical (and rather self-contradictory) view that this is a fact which cannot be expressed, or else find some legitimate way to affirm it.

The way forward is to recognize that truth and falling under do make sense globally, but as constructive, not classical, notions. In both cases we can recognize an indefinitely extensible quality: we are able to produce partial classical characterizations of truth and falling under, but any such characterization can be extended. This fits with the intuitionistic conception of mathematical reality as something which does not have a fixed global existence but instead is open-ended and can only be constructed in stages. The intuitionistic account may or may not be valid as a description of mathematics, but it unequivocally does capture the fundamental nature of truth and falling under. On pain of contradiction, these notions do not enjoy a global classical existence. However, they can indeed be built up in an open-ended sequence of stages.

The central concept in constructive mathematics is proof, not truth. And this is just the linguistic resource we need to make sense of second order quantification. Writing $\Box A$ for "A is provable" and $p \vdash A$ for "p proves A", we have

$$\Box A \leftrightarrow (\exists p)(p \vdash A).$$

Note that this expression can be universally quantified because no appearance of A is assertoric, so that it is a legitimate definition of the box operator. Note also that there is no question about formulating a global definition of the proof relation, as this is a primitive notion which we do not expect to define in any simpler terms. We can therefore, following the intuitionists, give a global constructive definition of truth by saying

$$A \text{ is constructively true} \leftrightarrow A \text{ is provable} \qquad (**)$$

and we can analogously give a global constructive definition of falling under by saying

$$x \text{ constructively falls under } C \leftrightarrow C(x) \text{ is provable.} \qquad (\dagger\dagger)$$

More generally, we can use provability to repair use/mention problems in expressions that quantify over all concepts. Such expressions can be interpreted constructively, and the self-referential capacity of global second order quantification makes it unreasonable to demand a classical interpretation. In particular, we can solve the problem raised at the beginning of this section. The way we say that any sentence can be given a truth definition is: for every sentence A there is a predicate T such that the assertion $T(A) \leftrightarrow A$ is provable. Again, no appearance of A or T is assertoric, so the quantification is legitimate. More substantially, we can affirm that for any language $\mathcal{L}$ there is a predicate $T_{\mathcal{L}}$ such that inserting any sentence of $\mathcal{L}$ in the template "$T_{\mathcal{L}}(\cdot) \leftrightarrow \cdot$" yields a provable assertion. Thus, there is a global constructive definition of truth, and there are local classical definitions of truth, but the global affirmation that these local classical definitions always exist is constructive.

We can make the same points about falling under; here too we have both local classical and global constructive options. The new feature in this setting is that no local classical definition of falling under can be used to make sense of sentences in which quantified concept variables are supposed to function predicatively. In order to handle this problem we require a globally applicable notion of falling under, which means that the classical option is unworkable. We have to adopt a constructive approach.

4.

The global constructive versions of truth and falling under are not obviously paradoxical because the biconditional $\Box A \leftrightarrow A$ is not tautological. We cannot simply assume that asserting A is equivalent to asserting that A is provable. The extent to which this law holds is a function of both the nature of provability and the constructive interpretation of implication. This issue is analyzed in [4] (see also [5]); we find that the law

(1) $A \rightarrow \Box A$

is valid but the converse inference of A from $\Box A$ is legitimate only as a deduction rule, not as an implication. Although the law $\Box A \rightarrow A$ is superficially plausible, its justification is in fact subtly circular.

The relation of $\Box$ to the standard logical constants, interpreted constructively, is also investigated in [4]; we find that the laws

(2) $\Box(A \wedge B) \leftrightarrow (\Box A \wedge \Box B)$

(3) $\Box(A \vee B) \leftrightarrow (\Box A \vee \Box B)$

(4) $\Box((\exists x)A) \leftrightarrow (\exists x)\Box A$

(5) $\Box((\forall x)A) \leftrightarrow (\forall x)\Box A$

(6) $\Box(A \rightarrow B) \rightarrow (\Box A \rightarrow \Box B)$

are all generally valid. There is no special law for negation; we take $\neg A$ to be an abbreviation of $A \rightarrow \bot$ where $\bot$ represents falsehood, so using (6) we can say, for instance,

$$\Box(\neg A) \leftrightarrow \Box(A \rightarrow \bot) \rightarrow (\Box A \rightarrow \Box\bot).$$

But $\Box(\neg A)$ is not provably equivalent to $\neg\Box A$ in general.

We can now present a formal system for reasoning about concepts that allows quantified concepts to predicate. The language is the language of set theory, augmented by the logical constant $\Box$. Formulas are built up in the usual way, with the one additional clause that if A is a formula then so is $\Box A$.

The variables are taken to range over concepts and $\in$ is read as "constructively falls under". Thus no appearance of a variable in any formula is assertoric and we can sensibly quantify over any variable in any formula. The system employs the usual axioms and deduction rules of an intuitionistic predicate calculus with equality, together with the axioms (1)–(6) above, the deduction rule which infers A from $\Box A$, the extensionality axiom

(7) $x = y \leftrightarrow (\forall u)(u \in x \leftrightarrow u \in y)$,

and the comprehension scheme

(8) $(\exists x)(\forall r)(r \in x \leftrightarrow \Box A)$

where x can be any variable and A can be any formula in which x does not appear freely. (In this scheme the variable r is fixed.) The motivation for the comprehension scheme is that any formula defines a concept (possibly with parameters, if A contains free variables besides r), and what it means to constructively fall under that concept is characterized by (††). This is why we need the ability to explicitly reference the notion of provability. The ex falso law can be justified in this setting by taking $\bot$ to stand for the assertion $(\forall x, y)(x \in y)$.

We call the formal system described in this section CC (Constructive Concepts). This is a "pure" concept system in the sense that there are no objects besides concepts. Alternatively, we could (say) take the natural numbers as given and write down a version of second order arithmetic in which the set variables are interpreted as concepts. From a predicative point of view a third order system, with number variables, set variables, and concept variables, would also be natural [3].

5.

The system CC accomodates global reasoning about concepts. For instance, using comprehension we can define the concept *concept which does not provably fall under itself*. Denoting this concept R, we have

$$r \in R \qquad \leftrightarrow \qquad \Box(r \notin r).$$

Assuming $R \notin R$ then yields $\Box(R \notin R)$ by axiom (1), which entails $R \in R$ by the definition of R. This shows that $R \notin R$ is contradictory, so we conclude $\neg(R \notin R)$. On the other hand, assuming $R \in R$ immediately yields $\Box(R \notin R)$; but since $R \in R$ also implies $\Box(R \in R)$, we infer $\Box\bot$. So we have $R \in R \to \Box\bot$. In the language of [4], the assertion $R \notin R$ is false and the assertion $R \in R$ is weakly false.

Thus, we can reason in CC about apparently paradoxical concepts and reach substantive conclusions. But no contradiction can be derived, as we will now show. (The proof of the following theorem is similar to the proof of Theorem 6.1 in [4].)

Theorem 5.1: *CC is consistent.*

Proof: We begin by adding countably many constants to the language of CC. Let $\mathcal{L}$ be the smallest language which contains the language of CC and which contains, for every formula A of $\mathcal{L}$ in which no variable other than r appears freely, a constant symbol c_A. Observe that $\mathcal{L}$ is countable.

We define the level $l(A)$ of a formula A of $\mathcal{L}$ as follows. The level of every atomic formula and every formula of the form $\Box A$ is 1. The level of $A \wedge B$, $A \vee B$, and $A \to B$ is $\max(l(A), l(B)) + 1$. The level of $(\forall x)A$ and $(\exists x)A$ is $l(A) + 1$.

Now we define a transfinite sequence of sets of sentences F_α. These can be thought of as the sentences which we have determined not to accept as true. The definition of F_α proceeds by induction on level. For each α the formula $\bot$ belongs to F_α; $c_B \in c_A$ belongs to F_α if $A(c_B)$ belongs to F_β for some $\beta < \alpha$; $c_A = c_{A'}$ belongs to F_α if for some c_B, one but not both of $A(c_B)$ and $A'(c_B)$ belongs to F_β for some $\beta < \alpha$; and $\Box A$ belongs to F_α

if A belongs to F_β for some $\beta < \alpha$. (Recall that the constants c_A are only defined for formulas A in which no variable other than r appears freely. So expressions like $A(c_B)$ are unambiguous.) For levels higher than 1, we apply the following rules. $A \wedge B$ belongs to F_α if either A or B belongs to F_α. $A \vee B$ belongs to F_α if both A and B belong to F_α. $(\forall x)A$ belongs to F_α if $A(c_B)$ belongs to F_α for some constant c_B, and $(\exists x)A$ belongs to F_α if $A(c_B)$ belongs to F_α for every constant c_B. (Observe here that if $(\forall x)A$ is a sentence then A can contain no free variables other than x, so again the expression $A(c_B)$ is unambiguous.) Finally, $A \to B$ belongs to F_α if there exists $\beta \leq \alpha$ such that B belongs to F_β but A does not belong to F_β.

Since the language $\mathcal{L}$ is countable and the sequence (F_α) is increasing, this sequence must stabilize at some countable stage α_0. It is obvious that $\perp$ belongs to F_{α_0}. The proof is completed by checking that the universal closure of no axiom of CC belongs to F_{α_0}, and that the set of formulas whose universal closure does not belong to F_{α_0} is stable under the deduction rules of CC. This is tedious but straightforward. □

6.

The system CC gives correct expression to Frege's idea of formalizing reasoning about arbitrary concepts. Frege was impeded by the fact that the global notion of falling under is inherently constructive; treating this notion as if it were classical is the fatal mistake which gives rise to Russell's paradox. We can locate the essential error in Frege's analysis not in his Basic Law V, or any of his other axioms, but rather in his use of a language whose cogency depends on a fictitious global classical notion of falling under.

Analyzing the proof theoretic strength of CC will show us the degree to which it is possible, as Frege hoped, to base mathematical reasoning on the pure logic of concepts. The result is disappointing. The simplicity of the consistency proof given in Theorem 5.1 already reveals that CC must be a very weak system. We now present two positive results which show how (conservative extensions of) CC can in a certain sense interpret more standard formal systems in which the box operator does not appear.

The relevant sense is the notion of *weak interpretation* introduced in [4]. We say that a theory $\mathcal{T}_2$ in which we are able to reason about provability weakly interprets another theory $\mathcal{T}_1$ in the same language minus the box operator if every theorem of $\mathcal{T}_1$ is a theorem of $\mathcal{T}_2$ with all boxes deleted. Observe that deleting all boxes in all theorems of CC yields an inconsistency: as we saw earlier, we can prove in CC the existence of a concept R

which satisfies both $\neg\neg(R \in R)$ and $R \in R \to \Box\bot$, and deleting the box in the second formula produces the contradictory conclusions $\neg\neg(R \in R)$ and $\neg(R \in R)$. Notwithstanding this phenomenon, no inconsistent theory can be weakly interpreted in CC. This is because $\bot$ is a theorem of every inconsistent theory, and weak interpretability would imply that $\Box^k\bot$ must be a theorem of CC for some value of k. Since CC implements the deduction rule which infers A from $\Box A$, this would then imply that $\bot$ is a theorem of CC, i.e., that CC is inconsistent.

The first system we consider, $\mathrm{Comp}(\mathrm{PF}_T) + \mathrm{D}$, was discussed in [1], where its consistency was proven. Here we show that the intuitionistic version of this system is weakly interpretable in an extension of CC by definitions.

$\mathrm{Comp}(\mathrm{PF}_T) + \mathrm{D}$ is a positive set theory. Its language is the ordinary language of set theory augmented by terms which are generated in the following way. Any variable is a term; if s and t are terms then $s \in t$ and $s = t$ are positive formulas; if A and B are positive formulas then $A \wedge B$, $A \vee B$, $(\forall x)A$, and $(\exists x)A$ are positive formulas; if A is a positive formula and x is a variable then $\{x : A(x)\}$ is a term whose variables are the free variables of A other than x. The system consists of the comprehension scheme

$$y \in \{x : A(x)\} \leftrightarrow A(y),$$

where A is a positive formula and x and y are variables, together with the axiom D which states

$$(\exists x, y)(x \neq y).$$

The desired conservative extension CC' of CC is obtained by recursively adding, for every formula A and variables x and y, the term $\{x : \Box A(x)\}$ (whose variables are the free variables of A other than x) together with the axiom

$$y \in \{x : \Box A(x)\} \leftrightarrow \Box A(y).$$

Say that a formula is *increasing* if no implication appears in the premise of any other implication. Note that since we take $\neg A$ to be an abbreviation of $A \to \bot$, this also means that an increasing formula cannot position a negation within the premise of any implication, nor can it contain the negation of any implication.

Observe that the axiom $y \in \{x : \Box A(x)\} \leftrightarrow \Box A(y)$ is increasing if A is positive, and the formula $(\exists x, y)(x = y \to \Box\bot)$, which is easily provable in

CC, is also increasing. Since removing all boxes from these formulas recovers the axioms of $\mathrm{Comp}(\mathrm{PF}_{\mathcal{T}}) + \mathrm{D}$, the following result is now a consequence of ([4], Corollary 7.3).

Theorem 6.1: *CC' weakly interprets intuitionistic* $\mathrm{Comp}(\mathrm{PF}_{\mathcal{T}}) + \mathrm{D}$.

It is interesting to note that the extensionality axiom of CC is not increasing, so that we cannot weakly interpret intuitionistic $\mathrm{Comp}(\mathrm{PF}_{\mathcal{T}}) + \mathrm{EXT} + \mathrm{D}$ in CC'. The latter theory is in fact inconsistent [1]. We can also show that a different extension of CC by definitions weakly interprets intuitionistic second order Peano arithmetic minus the induction axiom. The extension is defined by adding a constant symbol 0 which satisfies

$$r \in 0 \leftrightarrow \Box\bot,$$

a unary function symbol S which satisfies

$$r \in Sx \leftrightarrow \Box(r = x),$$

and a constant symbol ω which satisfies

$$r \in \omega \leftrightarrow \Box(\forall z)[(0 \in z \wedge (\forall x)(x \in z \to Sx \in z)) \to r \in z].$$

The following formulas are easily proven in the resulting extension CC'':

$0 \in \omega$;
$x \in \omega \to Sx \in \omega$;
$Sx = 0 \to \Box\bot$;
$Sx = Sy \to \Box(x = y)$;
$(0 \in z \wedge (\forall x)(x \in z \to Sx \in z)) \to (\forall y)(y \in \omega \to \Box(y \in z))$.

Since the first four of these formulas are increasing, the claimed result again follows from ([4], Corollary 7.3).

Theorem 6.2: *CC'' weakly interprets intuitionistic second order Peano arithmetic minus induction.*

Since the induction axiom is not increasing it has to be excluded from this result. Thus, although CC proves a version of full second order induction, it nonetheless appears to possess only meager number theoretic resources.

References

1. M. Forti and R. Hinnion, The consistency problem for positive comprehension principles, *J. Symbolic Logic* **54** (1989), 1401-1418.
2. A. Tarski, The semantic conception of truth, *Philosophy and Phenomenlological Research* **4** (1944), 13-47.
3. N. Weaver, Axiomatizing mathematical conceptualism in third order arithmetic, arXiv:0905.1675.
4. _____, The semantic conception of proof, arXiv:1112.6126.
5. _____, Kinds of concepts, arXiv:1112.6124.

PERFECT INFINITIES AND FINITE APPROXIMATION

Boris Zilber

Mathematical Insitute
University of Oxford
24-29 St Giles, Oxford, OX1 3LB, UK
zilber@maths.ox.ac.uk

In this chapter we present an analysis of uses of *infinity* in "applied mathematics", by which we mean mathematics as a tool for understanding the real world (whatever the latter means). This analysis is based on certain developments in Model Theory, and lessons and question related to these developments.

1. Introduction

Model theory occupies a special position in mathematics, with its aim from the very outset being to study real mathematical structures from a logical point of view and, more ambitiously, to use its unorthodox methods and approaches in search of solutions to problems in core mathematics. Model-theorists made an impact and gained experience and some deep insights in many areas of mathematics: number theory, various fields of algebra, algebraic geometry, real and complex geometry, the theory of differential equations, real and complex analysis, measure theory. The present author believes that model theory is well-equipped to launch an attack on some prolems of modern physics. This article, in particular, discusses what sort of problems and challenges of physics can be tackled model-theoretically. Another topic of the discussion, in our view intrinsically related to the first one, is the way mathematical infinities arise from finite structures, the concept of limit and its variations.

2. Continuity and its Alternatives

2.1.

The mathematically best developed form of actual infinity is the notion of a continuous line and a continuous space. As we all well know this concept did not look indisputable to the ancient Greeks and has only become "intuitively obvious" perhaps since Newton made it a part of his Physics. In fact, the assumption that Newtonian physics takes place in a Euclidean space (and in a smooth form) is a powerful axiom from which most of the physics follows. Modern physics has moved away from the assumption that the space is Euclidean to a space being a manifold, but it is still the same idea of continuity. This causes a lot of trouble, e.g. showing up in non-convergent integral expressions that are dealt with by various heuristic tricks having no justification in continuous mathematics (see e.g. a very interesting example of such a calculation in [1]). These days a considerable proportion of physicists believe that the assumption of a continuous universe is false. But the formulation of an alternative paradigm will have to wait at least till the solution of the problem of quantum gravity.

So, why continuity is so crucial? The answer lies in the practices of physics. Continuity organises the structure of the physical world and gives it a certain *regularity*, as opposed to a potential chaos. Indeed, if we assume that the trajectory of a particle is smooth we can predict its position in the (near) future based on the knowledge of the past. The alternative seems to be destroying any prospect of having a predictive theory at all.

2.2.

A few words about spaces as manifolds: these are patched together from standard canonical pieces of a Euclidean space, such as an interval of a real line, a cube in 3-space and the higher-dimensional versions of these (or their complex analogues). A defining feature of the construction is its high degree of **homogeneity**: *a manifold M looks the same in a small neighbourhood of any of its points.* This can be expressed in more rigorous terms by saying that *there is a local isomorphism between neighbourhoods of any two points of M.*

2.3.

In fact, an alternative to continuity does exist and is well-known in mathematics and increasingly being used in physics. This is based on a topology of a different kind, the Zariski topology, which is coarse and in general does not allow metrisation.

Zariski topology enters mathematics and physics in at least two possible ways. One is by formally restricting one's study to the context of algebraic varieties and schemes (thus allowing only charts which are zero-sets of systems of polynomial equations over an abstract field, and maps which are rational). This seems to be unreasonably restrictive to physics although methods developed by Grothendieck's school allow mimicking many notions of analysis in this context and calculate very delicate cohomological invariants that alternatively can be visualised in complex geometry. The other appearance of algebraic geometry comes via complex analysis and the study of complex manifolds. A crucial feature of complex functions is that differentiabily (even just once) implies a very strong form of smoothness – the function becomes analytic. This eventually entails that the behaviour of compact complex manifolds is very close to that of general algebraic varieties. A manifestation of this is the theorem by Riemann stating that any one-dimensional compact complex manifold can be realised as a complex algebraic curve. As a matter of fact this is a corollary of a stronger theorem by Riemann about compact real surfaces with a Riemannian metric (Riemann surfaces): every such surface can be identified with a complex algebraic curve with a metric induced by the metric on the complex numbers. These sort of connections with metric geometry led physicists to appreciate the relevance of algebraic geometry. Recall that Calabi-Yau manifolds which, according to string theory, underpin the structure of physics are objects of the same dual nature. By some definition (slightly stronger than the usual one, see [2]) they turn out to be algebraic.

2.4.

The shift from continuous geometry based on the reals towards algebraic geometry and an even more general algebraic and category-theoretic mathematical setting in physics is characteristic of our time. Moreover, there is a growing realisation of the need to reconsider the mathematical constructs at the foundations of physics. In [3] we read: "Indeed, there has always been a school of thought asserting that quantum theory itself needs to be radically changed/developed before it can be used in a fully coherent quantum theory of gravity. This iconoclastic stance has several roots, of which, for us, the most important is the use in the standard quantum formalism of certain critical mathematical ingredients that are taken for granted and yet which, we claim, implicitly assume certain properties of space and time. Such an a priori imposition of spatio-temporal concepts would be a major error if they turn out to be fundamentally incompatible with what is needed for a theory of quantum gravity. A prime example is the use of the continuum which, in this context, means the real and/or complex numbers."

We suggest a way to approach this issue by aiming at identifying what *logically perfect mathematical structures* should be. Once this is achieved, these should be taken as background structures for physics. A successful definition of logical perfection must entail a degree of regularity for structures enjoying the property that makes them classifiable enough to have a good mathematical theory and flexible enough to model physical systems.

The idea of having a perfect structure as a mathematical basis for physics is not very original. Certainly, the Euclidean space in the background for Newtonian physics is perfect enough a structure. An even more characteritic example is provided by the perfect spheres underlying Ptolemaic astronomy, which later led to the introduction of more sophisticated structures, epicycles, approximating the motion of planets quite well. These may look totally inadequate today but one must keep in mind that the approximation by epicycles is essentially Fourier analysis, fully respectable in modern physics.

We use a model-theoretic approach built around the analysis of interaction of a mathematical structure and its description in a formal language (C.Isham and A.Döring in [3] start by discussing the type of language that can lie at the foundations of physics).

3. In Search of Logically Perfect Structures

3.1.

The main developments in model theory in recent decades have been centered around stability theory and the core of stability theory is the theory of categoricity in uncountable cardinals.

It is well-known that the first-order description of a structure $\mathbf{M}$ can be (absolutely) categorical if and only if $\mathbf{M}$ is finite, which is a quite trivial situation (unless we put a restriction on the size of the first order axiomatisation), hence the need for a more subtle definition.

A structure $\mathbf{M}$ is said to be categorical in cardinality λ if there is exactly one, up to isomorphism, structure $\mathbf{M}$ of cardinality λ satisfying the (first-order) theory $\mathrm{Th}(\mathbf{M})$.

In other words, if we add to Th(**M**) the (non first-order) statement that the cardinality of its universe is λ the description becomes categorical. Of special interest is the case of uncountable cardinality λ. In his seminal work [4] on categoricity M.Morley proved that the categoricity of a theory in one uncountable λ implies the categoricity in all uncountable cardinalities, so in fact the actual value of λ does not matter. What we have is a large structure (of an uncountable cardinality) which has a concise (countable) categorical description.

There are purely mathematical arguments towards accepting the above for a definition of perfection. First, we note that the theory of the field of complex numbers (in fact any algebraically closed field) is uncountably categorical. So, the field of complex numbers is a perfect structure, and so are all objects of complex algebraic geometry by virtue of being definable in the field.

It is also remarkable that Morley's theory of categoricity (and its extensions) exhibits strong regularities in models of categorical theories generally. First, the models have to be highly *homogeneous*, in the sense technically different from one discussed for manifolds in subsection 2.2 but similar in spirit (in fact, it follows from results of complex geometry that any compact complex manifold is ω-stable of finite Morley rank; many such manifolds are categorical). Moreover, a notion of *dimension* (the Morley rank) is applicable to definable subsets in categorical structures, which gives one a strong sense of working with curves, surfaces and so on in this very abstract setting. A theorem of the present author states more precisely that an uncountably categorical structure M is either reducible to a 2-dimensional "pseudo-plane" with at least a 2-dimensional family of curves on it (so is non-linear), or is reducible to a linear structure like an (infinite-dimensional) vector space, or to a simpler structure like a G-set for a discrete group G. This led to a Trichotomy conjecture, [9], which specifies that the non-linear case is reducible to algebraically closed fields, that effectively implies that M in this case is an object of algebraic geometry over an algebraically closed field.

There remains the question of whether the restriction to the first-order language is natural. Although we would like to keep this open there are good reasons that the ultimate logical perfection must be first-order. The use of first-order languages was effectively suggested by Hilbert for reasons of its finitarity, the ability (and the restriction) to use only expressions of finite length, which agrees well with practicalities of physics.

Alternatively, we could extend the idea to dealing with very large finite structures as if they are infinite with (uncountably) categorical theories. This could be formalised provided a right notion of approximation is found. We develop this approach in the second part of the chapter.

3.2.

The arguments above sugest that because of the high degree of logical perfection exhibited by uncountably categorical structures they must already be in the centre of mathematics. Certainly, abstract mathematics being based on pure logic finds a special interest in objects having a concise and complete description. Physics, supposedly based on objective reality, is different in this regard but many recent discussions about the possible interaction between human intelligence and the structure of physical reality as we perceive it (anthropic principle) may suggest a similar relevance of the notion of categoricity to physics. In particular, the notion of *algorithmic compressibility* (the idea and the term comes from [5]) seems to be in close relation to categoricity as expression for "concise and complete". In [6] we read that "the existence of regularities [in the real world] may be expressed by saying that the world is algorithmically compressible." And further on, "The fundamental laws of physics seem to be expressible as succinct mathematical statements. ... does this fact tell us something important about the structure of the brain, or the physical world, or both?".

3.3.

Although we now know that the Trichotomy conjecture is technically false, it is believed to be (in words of D.Marker) "morally true". The significance and the limits of the conjecture are now understood much better.

E.Hrushovski showed that the Trichotomy conjecture in its full generality is false, producing a series of counter-examples, beginning in [8]. Since then other variations of counter-examples appeared exhibiting various possibilities of how the Trichotomy conjecture may fail. Remarkably, after more than 20 years since [8] Hrushovski's construction is the only source of counterexamples. So, what does this construction demonstrate – the failure of the philosophy behind the conjecture and existence of chaotic, pathological structures consistent with categoricity, or incompleteness of the conjecture itself?

In [10] this author showed that Hrushovski's construction when applied in the right mathematical context produces in fact quite perfect structures. The structure in [10] is an analogue of the field of complex numbers with exponentiation, $(\mathbb{C}, +, \cdot, \exp)$. It turns out that the $L_{\omega_1,\omega}(Q)$-theory of this structure (call it a *perfect exponentiation*) is categorical in uncountable cardinalities, so there is exactly one, up to isomorphism, such structure of cardinality continuum. On the other hand, all comparisons between the structure with perfect exponentiation and the actual complex numbers with exponentiations suggest that the one with perfect exponentiation may be isomorphic to the genuine one. This suggestion taken as a hypothesis has a number of important consequences, including the remarkable Schanuel conjecture that, extended appropriately, covers practically all which is known or conjectured in the transcendental number theory.

Today this pattern of linking Hrushovski's counter-examples to classical analytic structures (based on classical transcendental functions) is supported by more case studies including the study of the Weierstrass $\mathfrak{P}$-function and the modular function j (the j-invariant for elliptic curves). So, one may suggest that although the Trichotomy version of the conjecture is false, there is another, more credible mathematical interpretation of the general principle of "logically perfect structures" with algebraic geometry replaced by its appropriate analytic extension.

In any case there is a feeling that something serious is going on around the principle of logical perfection. The above mentioned categoricity theorem for perfect exponentiation is not trivial. Its proof requires a considerable input from model theory (mainly Shelah's theory of *excellence* for *abstract elementary classes*, [11]), but also a serious amount of results from transcendental and Diophantine number theory. A sense of magic is present when in order to prove the categoricity you identify the need of a number-theoretic fact not known to you prior to this work and you learn from experts that the fact indeed holds, but has been proven just a few years ago. In [12] we proved the equivalence of categoricity for some type of structures (universal covers of abelian varieites) to a conjunction of number-theoretic statements (among these the so called "hard theorem of Serre"). Most of the statements are known, either as facts or as conjectures. Some have been proved just recently.

To sum up this line of developments around categoricity we would like to stress a remarkable phenomenon. The assumption of categoricity led one to a construction (of perfect exponentiation) that is not based on continuity but nevertheless turned out to possess all the features observed for a classical analytic transcendental function and, moreover, predicted other properties that have been independently conjectured.

3.4. Topological structures

The crucial, in this author's view, improvement to the notion of logical perfection has been introduced in [13] by Hrushovski and this author (a similar but weaker notion was already present in the paper [14] by Pillay and Srour). The idea is to account for a topological ingredient in logic, essentially by giving special significance to *positive* formulas, assuming that in the given structure positively definable sets give rise to a (coarse) topology. In [15] and in the second part of this chapter such structures are called *topological structures*.

Of course, the syntax of an axiomatisation has always been of importance in logic and must have been part of any notion of logical perfection, but in model theory of the 1960 it was found convenient to abstract away from the syntax of formulas to the framework of Boolean algebras of definable sets. This now needs a correction, especially if one approaches the subject with a view of applications in physics. Clearly, a formulation of a physical low is expected to have the form of an equation rather than its negation. To even start thinking about the idea of *approximation* one needs to assume the possibility that certain type of statement, e.g. equations, can happen to fail in reality but be considered true in approximation.

Now looking back at our examples one can rephrase that model theory of algebraically closed fields becomes algebraic geometry if we pay special attention to positively definable sets, i.e. the sets closed in Zariski topology. This motivated the terminology in [13] where we called a 1-dimensional categorical topological structure a Zariski structure provided the topological dimension agrees with Morley rank and the topology satisfies certain "dimension theorem" which holds in algebaic geometry on smooth varieties: in an n-dimensional space M every irreducible component of the intersection of two closed irreducible sets S_1 and S_2 is of dimension at least $\dim S_1 + \dim S_2 - n$ (we call the latter property *presmoothness* now).

In [15] this definition has been lifted to aribitrary dimensions. Note also that the Zariski topology in [13] and generally in first-order categorical structures where it can be defined is Noetherian. The field with perfect exponentiation of subsection 3.3 can also be treated as a topological structure with a Zariski-type topology which is not Noetherian, and some of key questions about the topology on this field remain unanswered.

3.5.

The arguments above suggest a choice for the notion we have been looking for. From now on logically perfect structures will be identified as (Noetherian) Zariski structures.

This agrees with the hierarchy of classes of structures (theories) developed by Shelah's classification theory. Zariski structures rightly can be placed in the centre of the classification picture surrounded by "less perfect" classes, such as classes of ω-stable, superstable, stable theories and so on. The theory of formally real fields has its place in the classification outside the class of stable theories but not very far from it.

Does this classification indicate an order of "importance" of mathematical structures? Certainly not. Real analysis on complex algebraic varieties provides an invaluable insight in the mathematics of purely algebraically defined object. Yet, as far as physics is concerned, there may be good reasons to see certain structures (or certain choice of languages) as basic, and other structures as auxiliary. This would agree with now broadly accepted Heisenberg's programme of basing the theory of quantum physics on "the relationships between magnitudes that are in principle observable".

3.6.

By reducing our analysis of logical perfection to Zariski geometries we achieve at least one meaningful gain. Firstly, this class is rich in mathematically significant examples, e.g. compact complex manifolds in their natural language are Zariski geometries. And secondly, this class allows a fine classification theory and, in particular, essentially satisfies the Trichotomy principle (that is the Trichotomy conjecture within the class is proven to be true).

The following is the main classification result by Hrushovski and this author [13]

Theorem. *For a non-linear Zariski geometry* **M** *there is an algebraically closed field* F *and a nonconstant Zariski-continuous "meromorphic" function* $\psi : \mathbf{M} \to \mathrm{F}$.

In particular, if $\dim M = 1$ *then there is a smooth algebraic curve* C_M *and a Zariski-continuous finite covering map*

$$p : \mathbf{M} \to C_M(\mathrm{F}),$$

the image of any relation on $\mathbf{M}$ *is just an algebraic relation on* C_M.

The remarkable message of this theorem is that the purely logical criteria that lead to the definition of Zariski geometries materialise in an, in fact, uniquely defined by $\mathbf{M}$ algebraically closed field F. The proof effectively reconstructs main ingredients of algebraic geometry starting from the most abstract ones and leading consequently to the reconstruction of the field itself. Note that the linear case which is not covered by the theorem, is reasonably classifiable, although the full account of this case has not been given yet.

The word "essentially" in the reference to the Trichotomy principle above is to indicate that the reduction to algebraic geometry is not as straightforward as one imagined when Zariski geometries were first introduced. Indeed, there is an example of *non-classical* Zariski geometries $\mathbf{M}$ of dimension 1 which is not reducible to an algebraic curve, but is only "finite over" the algebraic curve C_M. In other words, $\mathbf{M}$ can be obtained by "inserting" a finite structure over each point of C_M in some uniform way, but so that the construction is not reducible to a direct product in any sensible form.

Note that one-dimensional Zariski geometries that originate from compact complex manifolds are classical also in the sense above due to the classification of compact Riemann surfaces; they all can be identified as complex algebraic curves, that is $F = \mathbb{C}$ and the covering map $p : \mathbf{M} \to C_M(F)$ is the identity.

3.7.

The defect of non-classicality seemed insignificant in the beginning, partly because it did not affect a number of applications that followed and partly because "finite" sounds almost as "trivial" in model theory.

But the situation is much more interesting if one tries to understand the non-classical examples from the geometer's and even the physicst's point of view.

The most comprehensive modern notion of a *geometry* is based on the consideration of a *co-ordinate algebra* of the geometric object. The classical meaning of a coordinate algebra comes from the algebra of *co-ordinate functions* on the object, that is functions $\psi : \mathbf{M} \to F$ as in subsection 3.6, of a certain class. The most natural algebra of functions for Zariski geometries seems to be the algebra $F[\mathbf{M}]$ of Zariski-continuous functions. But in a non-classical case by virtue of construction $F[\mathbf{M}]$ is naturally isomorphic to $F[C_M]$, the algebra of Zariski-continuous (definable) functions on the algebraic curve C_M. That is the only geometry which we see by looking into $F[\mathbf{M}]$ is the geometry of the algebraic curve C_M.

In [16], in order to see the rest of the structure we extended $F[\mathbf{M}]$ by introducing auxiliary *semi-definable* functions, which satisfy certain *equations* but are not uniquely defined by these equations. The F-algebra $\mathcal{H}(\mathbf{M})$ of semi-definable functions contains the necessary information about $\mathbf{M}$ but is not canonically defined. On the other hand it is possible to define an F-algebra $\mathcal{A}(\mathbf{M})$ of linear operators on the linear space $\mathcal{H}(\mathbf{M})$ in a canonical way, depending on $\mathbf{M}$ only. Moreover, using a specific auxiliary function one can define a natural involutive mapping $X \to X^*$ on generators of $\mathcal{A}(\mathbf{M})$ thus defining a weak version of adjoints. We wrote down explicit lists of generators and defining relations for algebras $\mathcal{A}(\mathbf{M})$ for some examples and demonstrated that $\mathcal{A}(\mathbf{M})$ as an abstract algebra with involution contains all the information needed to recover the "hidden" part of structure $\mathbf{M}$.

Later studies in [17], [18] and yet unpublished theses of V.Solanki and D.Sustretov confirm that this is a typical situation. There are lessons that one learns from it:

- The class of Zariski geometries extends algebraic geometry over algebraically closed fields into the domain of non-commutative, quantum geometry.
- For large classes of quantum algebras Zariski geometries serve as counterparts in the duality "co-ordinate algebra – geometric object" extending the canonical duality of commutative geometry.
- The non-commutative co-ordinate algebras for Zariski geometries emerge essentially for the same reasons as they did in quantum physics.

3.8.

In [16] one more important observation was made. The examples of non-classical Zariski geometries come in uniform families with variations within a family given by the size of the fibre of the covering map p. It is natural to ask what can be seen when the size of the fibre tends to infinity. Is there a well-defined limit structure? For examples studied in [16] it is possible to introduce a discrete metric on the Zariski structures so that there is a well-defined limit structure that was identified as a real differential manifold with a non-trivial gauge field on it.

This example demonstrates that one can study non-classical Zariski objects seeing them "from afar" in quite a classical way, as real metric, even differentiable manifolds. This agrees with the general principle of physics of how quantised theories must behave: *when quantum mechanics is applied to big structures, it must give the results of classical mechanics.*

Along with this the question arises, *what is the appropriate version of approximation (limit)?* The one used in [16] is the Hausdorff limit for sequences of metric space. But to apply the Hausdorff (or Gromov-Hausdorff) limit one needs that the quantum structures already have a natural metric on them. Is it possible that a sequence of non-metrisable structures have a limit structure with a nice metric?

The questions above have also practical significance. As a matter of fact the structures in finite fibres hold key information, in essense they are finite-dimensional representations of the operator algebras involved, see [17]. The problem of limit to a large degree amounts to a choice of discrete (even finite) models of physical processes in question. Providing a solution to this problem one possibly will be solving problems with non-convergent calculations mentioned in subsection 2.1.

In the second part of this chapter we introduce a notion of a *structural approximation* that we believe has a potential to give a mathematically rigorous formalisation to often heuristic approximation procedures used by physicists. (A beautiful and honest account by a mathematician attemting to translate the physicist's vision of "matrix algebras converging to the sphere" into a mathematical concept provides the introductory section of [20].)

The notion of structural approximation is closely linked to the idea of treating structures as topological, in topology induced by the choice of the language, see subsection 3.4. The example in [16] mentioned above demonstrate that there exists a possibility that beautiful and mathematically rich differential-geometric structures of physics could be just limits of discrete logically perfect ones. We prove in subsection 6.1 that algebraically closed fields, that shape the "classical part" of a logically perfect structure (see subsection 3.6), are approximable by finite fields. So one can even make a suggestion that structures central to physics have "perfect" finite approximations, or even that they **are** finite. In this regard it is worth mentioning an attempt to build a physical theory based on a large finite field instead of the reals (P. Kustaanheimo and others). A non-trivial argument in subsection 6.1 proves that this wouldn't be possible. The complex numbers is the only locally compact field that can be approximated by finite fields preserving structural properties. One can say that if the physical world is indeed parametrised by a huge finite field, then it would look from afar as an object of complex algebraic geometry. Which agrees well with trends in modern mathematical physics.

In the last subsection of the second part we discuss an open problem regarding structural approximation of compact Lie groups, such as SO(3), by finite groups. We provide some references in literature that links this problem to key issues in quantum field theory.

4. Structural Approximation

4.1. *Topological structures*

Following [15] we consider structures $\mathbf{M}$ endowed with a topology in a language $\mathcal{C}$. We say that the language is topological meaning that it is a relational language which will be interpreted so that any n-ary $P \in \mathcal{C}$ (basic $\mathcal{C}$-predicate) defines a closed subset $P(\mathbf{M})$ of M^n in the sense of a topology on M^n, all $n \in \mathbb{N}$. Not every closed subset of the topology in question is necessarily assumed to have the form $P(\mathbf{M})$, so those which are will be called $\mathcal{C}$-**closed**.

We assume that the equality is closed and all structures in question satisfy the $\mathcal{C}$-theory which ascertains that

- if $S_i \in \mathcal{C}$, $i = 1, 2$, then $S_1 \& S_2 \equiv P_1$, $S_1 \vee S_2 \equiv P_2$, for $P_1, P_2 \in \mathcal{C}$;
- if $S \in \mathcal{C}$, then $\forall x S \equiv P$, for some $P \in \mathcal{C}$.

We say that a $\mathcal{C}$-structure $\mathbf{M}$ is **complete** if, for each $S(x, y) \in \mathcal{C}$ there is $P(y) \in \mathcal{C}$ such that $\mathbf{M} \vDash \exists x S \equiv P$.

Note that we can always make $\mathbf{M} = (M, \mathcal{C})$ complete by extending $\mathcal{C}$ with relations P_S corresponding to $\exists x S$ for all S in the original $\mathcal{C}$. We will call such an extension of the topology **the formal completion** of $\mathbf{M}$.

We say $\mathbf{M}$ is **quasi-compact** (often just **compact**) if $\mathbf{M}$ is complete, every point in M is closed and for any filter of closed subsets of M^n the intersection is nonempty.

Remark 4.1: The family of $\mathcal{C}$-closed sets forms a basis of a topology, the closed sets of which are just the infinite interestions of filters of $\mathcal{C}$-closed sets (the topology generated by $\mathcal{C}$).

If the topology generated by $\mathcal{C}$ is Noetherian then its closed sets are exactly the ones which are $\mathcal{C}$-closed.

4.2. Structural approximation

Definition 4.2: Given a structure $\mathbf{M}$ in a topological language $\mathcal{C}$ and structures $\mathbf{M}_i$ in the same language we say that $\mathbf{M}$ is **approximated** by a family $\{\mathbf{M}_i = \mathbf{M} : i \in I\}$ along an ultrafilter D on I if, for some elementary extension $\mathbf{M}^D \succcurlyeq \prod_D \mathbf{M}_i$, there is a surjective homomorphism

$$\lim : \mathbf{M}^D \to \mathbf{M}.$$

Proposition 4.3: *Suppose every point of $\mathbf{M}$ is closed and $\mathbf{M}$ is approximated by the sequence $\{\mathbf{M}_i = \mathbf{M} : i \in I\}$ for some I along an ultrafilter D on I, such that $\mathbf{M}^D$ is saturated. Then the formal completion of $\mathbf{M}$ is quasi-compact.*

Proof: Consider the $\mathbf{M}_i$ and $\mathbf{M}$ formally completed, that is in the extended toplogy. Note that the given $\lim : \mathbf{M}^D \to \mathbf{M}$ is still a homomorphism in this language, since a homomorphism preserves positive formulas.

Closedness of points means that for every $a \in \mathbf{M}$ there is a positive one-variable $\mathcal{C}$-formula P_a with the only realisation a in $\mathbf{M}$. Under the assumptions for $^*\mathbf{M} \succ \mathbf{M}$, setting for $a \in \mathbf{M}$, $i(a)$ to be the unique realisation $\hat{a} \in {}^*\mathbf{M}$ of P_a we get an elementary embedding $i : \mathbf{M} \prec {}^*\mathbf{M}$. Now $\lim$ becomes a specialisation onto $\mathbf{M}$. This implies by [22] (see also a proof in [15]) that $\mathbf{M}$ is quasi-compact.   □

In agreement with the proposition we will consider only approximations to quasi-compact structures.

Proposition 4.4: *Suppose $\mathbf{M}$ is a quasi-compact topological $\mathcal{C}$-structure and $\mathbf{N}$ is an $|M|$-saturated $\mathcal{C}$-structure such that $\mathbf{N}$ is complete and for every positive $\mathcal{C}$-sentence σ*

$$\mathbf{N} \vDash \sigma \Rightarrow \mathbf{M} \vDash \sigma.$$

Then there is a surjective homomorphism $\lim : \mathbf{N} \to \mathbf{M}$.

Proof: Given $A \subseteq N$, a partial *strong homomorphism* $\lim_A : A \to \mathbf{M}$ is a map defined on A such that for every $a \in A^k$, $\hat{a} = \lim_A a$ and $S(x, y) \in \mathcal{C}$ such that $\mathbf{N} \vDash \exists y S(a, y)$, we have $\mathbf{M} \vDash \exists y\, S(\hat{a}, y)$.

When $A = \emptyset$ the map is assumed empty but the condition still holds, for any sentence of the form $\exists y\, S(y)$. So it follows from our assumptions that $\lim_\emptyset$ does exist.

Claim 1: *Suppose for some $A \subseteq N$ there is a partial strong homomorphism* $\lim_A : A \to \mathbf{M}$, *and $b \in N$. Then* $\lim_A$ *can be extended to a partial strong homomorphism* $\lim_{Ab} : Ab \to \mathbf{M}$.

Proof: Let $\mathbf{N} \models \exists z\, S(a, b, z)$, for $S(x, y, z)$ a positive formula and a a tuple in $\mathbf{N}$. Then $\mathbf{N} \models \exists yz\, S(a, y, z))$ and hence $\mathbf{M} \models \exists yz S(\hat{a}, y, z))$.

It follows that the family of closed sets in $\mathbf{M}$ defined by $\{\exists z S(\hat{a}, y, z) : \mathbf{N} \models \exists z\, S(a, b, z)\}$ is a filter. By quasi-compactness of $\mathbf{M}$ there is a point, say $\hat{b}$ in the intersection. Clearly, letting $\lim_{Ab} : b \to \hat{b}$, we preserve formulas of the form $\exists z\, S(x, y, z)$. Claim proved.

Claim 2: *For $A \subset N$, $|A| \leq |M|$, assume $\lim_A$ exists and let $\hat{b} \in M \setminus A$. Then there is a $b \in N$ and an extension* $\lim_{Ab} : b \mapsto \hat{b}$.

Proof: Consider the type over A,

$$p = \{\neg \exists z\, S(a, y, z) : \mathbf{M} \models \neg \exists z\, S(\hat{a}, \hat{b}, z) : \hat{a} = \lim_A a, a \subset A, S \in \mathcal{C}\}.$$

This is consistent in $\mathbf{N}$ since otherwise

$$\mathbf{N} \models \forall y \bigvee_{i=1}^{k} \exists z_i\, S_i(a, y, z_i)$$

for some finite subset of the type. The formula on the right is equivalent to $P(a)$, some $P \in \mathcal{C}$, so

$$\mathbf{M} \models \forall y \bigvee_{i=1}^{k} \exists z_i\, S_i(\hat{a}, y, z_i)$$

$$\mathbf{M} \models \bigvee_{i=1}^{k} \exists z_i\, S_i(\hat{a}, \hat{b}, z_i),$$

the contradiction. Claim proved.

To finish the proof of the proposition consider a maximal partial strong homomorphism $\lim = \lim_A : A \to M$. By Claim 1, $A = N$, so $\lim$ is a total map on N. By Claim 2, $\lim$ is surjective. □

Theorem 4.5: *Let $\mathbf{M}_i$, $i \in I$, be a family of formally completed topological $\mathcal{C}$-structures. Let $\mathbf{M}$ be a formally complete quasi-compact $\mathcal{C}$-structure. Then the following two conditions are equivalent,*

(i) *there is an unltrafilter D on I such that* $\lim_D \mathbf{M}_i = \mathbf{M}$,
(ii) *for every sentence $P \in \mathcal{C}$ such that $\mathbf{M} \models \neg P$ there is an $i \in I$,* $\mathbf{M}_i \models \neg P$.

Proof: (i) implies (ii) since positive formulas are preserved by homomorphisms.

We now prove (ii)⇒(i). For a given sentence $P \in \mathcal{C}$ let

$$I_P = \{i \in I : \mathbf{M}_i \models \neg P\}.$$

Let

$$D_{\mathbf{M}} = \{I_P : \mathbf{M} \models \neg P\}.$$

$D_{\mathbf{M}}$ is a filter. Indeed, every element of $D_{\mathbf{M}}$ is nonempty by (ii). Also, the intersection of two elements of $D_{\mathbf{M}}$ is an element of $D_{\mathbf{M}}$, since $P_1 \vee P_2 \equiv P \in \mathcal{C}$, for any $P_1, P_2 \in \mathcal{C}$, by definition.

Take D to be any ultrafilter on I extending $D_{\mathbf{M}}$. The statement follows from Proposition 4.4. □

5. Examples

In this section we assume for simplicity that $\mathbf{M}^D = \prod \mathbf{M}_i / D$.

5.1. *Metric spaces*

Let $\mathbf{M}$ and $\mathbf{M}_i$ be metric spaces in the language of binary predicates $d_r^{\leq}(x, y)$ and $d_r^{\geq}(x, y)$, all $r \in \mathbb{Q}$, $r \geq 0$, with the interpretation $\mathrm{dist}(x, y) \leq r$ and $\mathrm{dist}(x, y) \geq r$ correspondingly. The sets given by positive existential formulas in this language form our class $\mathcal{C}$.

Proposition 5.1: *Assume* $\mathbf{M}$ *is compact and*

$$\mathbf{M} = GH\text{-}\lim_{D} \mathbf{M}_i,$$

the Gromov-Hausdorff limit of metric spaces along a non-principal ultrafilter D on I. Then

$$\mathbf{M} = \lim_{D} \mathbf{M}_i.$$

Proof: By definition, for any n there is an $X_n \in D$ such that $\mathrm{dist}(M_i, M) \leq \frac{1}{n}$, in a space containing both all the M_i for $i \in X_n$ and M. For any $\alpha \in \prod_i M_i$ define $\hat{\alpha}$ to be an element of M^I such that $\hat{\alpha}(i)$ is an element of M at a minimal distance from $\alpha(i)$ (choose one if there is more than one at the minimal distance). Let a_α be the limit point of the sequence $\{\hat{\alpha}(i) : i \in I\}$ along D in M. We define

$$\lim_{D} \alpha := a_\alpha.$$

It follows from the construction that, for $\alpha, \beta \in \prod_i M_i$,

$$\{i \in I : M_i \vDash d_r(\alpha(i), \beta(i))\} \in D \Rightarrow M \vDash d_r(\lim_D \alpha, \lim_D \beta). \qquad \square$$

5.2. Cyclic groups in profinite topology

Consider the coset-topology on $\mathbb{Z}$ and $\mathbb{Z}/n\mathbb{Z}$. The compactification of $\mathbb{Z}$ is then $\hat{\mathbb{Z}}$, the profinite completion. Choose a non-principal ultrafilter D on $\mathbb{N}$ so that $m\mathbb{N} \in D$ for every positive integer m (a *profinite ultrafilter*).

    Claim.

$$\prod_D \mathbb{Z}/n\mathbb{Z} \cong \hat{\mathbb{Z}} \dotplus \mathbb{Q}^\kappa \dotplus T, \textit{ some cardinal } \kappa \textit{ and the torsion subgroup } T. \quad (5.1)$$

Proof: Follows from the Eklof-Fisher classification of saturated models of Abelian groups [23]. $\qquad \square$

Proposition 5.2: *The group $\hat{\mathbb{Z}}$ is approximated by $\mathbb{Z}/n\mathbb{Z}$ in the profinite topology. That is there is a surjective homomorphism*

$$\lim : \prod_D \mathbb{Z}/n\mathbb{Z} \to \hat{\mathbb{Z}}.$$

Proof: Define $\lim : \hat{\mathbb{Z}} \dotplus \mathbb{Q}^\kappa \dotplus T \to \hat{\mathbb{Z}}$ to be the projection (with kernel $\mathbb{Q}^\kappa \dotplus T$). $\qquad \square$

    As an example, consider the element (sequence) $\gamma(n)$ such that $\gamma(n) = \frac{n}{2} \bmod n$, all $n \in 2\mathbb{N}$. Then $2\gamma = 0$ in $\prod_D \mathbb{Z}/n\mathbb{Z}$, a torsion element, so $\lim \gamma = 0$.

5.3. The ring of p-adic integers

Consider the sequence of finite rings $\mathbb{Z}/p^n\mathbb{Z}$ and its ultraproduct

$$R_p := \prod_D \mathbb{Z}/p^n\mathbb{Z}$$

over a non-principal ultrafilter. Let $J_p \subset R_p$ be the ideal of divisible elements, that is the maximal ideal with the property $kJ_p = J_p$ for every integer k.

    Claim. R_p/J_p is an integral domain.

    Indeed, $a \cdot b \in J_p$ if and only if $a \cdot b$ is p^∞-divisible, if and only if a or b is p^∞-divisible, if and only if $a \in J_p$ or $b \in J_p$. Claim proved.

Introduce a metric on $\bar{R}_p = R_p/J_p$ setting the distance $d(a, b) \leq p^{-k}$ if $a - b \in p^k \bar{R}_p$. Then

$$d(a, b) \leq p^{-k} \text{ for all } k \text{ iff } a - b \text{ is } p^\infty\text{-divisible iff } a = b.$$

Clearly, the diameter of the metric space is 1 and it follows (using the saturatedness of R_p) that $\bar{R}_p$ is compact in the corresponding topology.

It follows that

$$R_p/J_p \cong \mathbb{Z}_p,$$

the quotient is isomorphic to the ring of p-adic integers.

We thus have proved

Proposition 5.3: *The ring $\mathbb{Z}_p$ of p-adic integers is approximated by the finite rings $\mathbb{Z}/p^n\mathbb{Z}$. That is there is a surejective homomorphism of rings*

$$\lim_D : \prod \mathbb{Z}/p^n\mathbb{Z} \to \mathbb{Z}_p.$$

5.4. *Compactified groups*

Call a compact topological structure **M** a **compactified group** if there is a closed subset $P \subset M^3$ and an open dense subset $G \subset M$ such that the restriction of P to G is a graph of a group operation on G, $P(g_1, g_2, g_3) \equiv g_1 \cdot g_2 = g_3$, and $P \cap G \times M^2$ defines an action of G on M.

We usually write such an **M** as $\bar{G}$.

Example 5.4: The structure $\check{\mathbb{Z}} = \mathbb{Z} \cup \{-\infty, +\infty\}$ with the ternary relation $S(x, y, z)$, defined as the closure of the graph of addition in the metric of the real line (the two-point compactification of $\mathbb{Z}$). By this definition $\models \forall z\, S(-\infty, +\infty, z)$, $\models \forall z\, S(z, +\infty, +\infty)$ and $\models \forall z\, S(z, -\infty, -\infty)$.

Example 5.5: The projective space $\mathbf{P}^{n^2}(\mathrm{F})$, for F an algebraically closed field, is a compactified group $\mathrm{GL}_n(\mathrm{F})$ in the Zariski topology of the projective space.

Example 5.6: The projective space $\mathbf{P}^n(\mathrm{F})$, for F an algebraically closed field, is a compactified additive group F^n, in the Zariski topology of the projective space, and in the metric topology if $\mathrm{F} = \mathbb{C}$.

In particular, for $n = 1$, Example 5.5 is a 2-point compactification of the multiplicative group of the fields, and Example 5.6 is a 1-point compactification of the additive group.

5.5. *Cyclic groups in metric topology and their compactifications*

The compactification of $\mathbb{Z}$ in the metric topology corresponding to the usual embedding of integers into the Riemann sphere $\mathbf{P}^1(\mathbb{C})$ is obviously $\bar{\mathbb{Z}} = \mathbb{Z} \cup \{\infty\}$, with the addition relation $S(x, y, z)$ (see Example 5.4) extended to the extra element: $\models \forall x\, S(x, \infty, \infty)$ and $\models \forall z\, S(\infty, \infty, z)$. We still write $x + y = z$ instead of $S(x, y, z)$.

For a finite cyclic group $\mathbb{Z}/n\mathbb{Z}$ define a metric as the metric of the regular n-gon with side 1 on the plane, induced by the metric of the plane.

We identify elements of $\prod_D \mathbb{Z}/n\mathbb{Z}$ with sequences $a = \{a(n) \in \mathbb{Z}/n\mathbb{Z} : n \in \mathbb{N}\}$ modulo D, any given non-principal ultrafilter.

Define

$$\lim a = \begin{cases} m, & \text{if } \{n \in \mathbb{N} : a(n) = m + n\mathbb{Z}\} \in D \\ \infty, & \text{otherwise.} \end{cases}$$

In other words, all bounded elements of $\prod_D \mathbb{Z}/n\mathbb{Z}$, which have to be eventually constant, specialise to their eventual value, and the rest go into ∞.

This is a surjective homomorphism onto $\bar{\mathbb{Z}}$ in the language $\{S\}$ and so lim also preserves (the topology of) the positively definable subsets. But lim is not a homomorphism in the language for metric, since for an unbounded element $a \in \prod_D \mathbb{Z}/n\mathbb{Z}$ we have $\lim a = \infty = \lim(a + 1)$ but $\models d_{\bar{1}}^{\geq}(a, a+1)$ while $\models \neg d_{\bar{1}}^{\geq}(\infty, \infty)$.

The downside of the 1-point compactification is that $\bar{\mathbb{Z}}$ "believes" that every its element is divisible, that is

$$\bar{\mathbb{Z}} \models \forall x \exists y\, x = \underbrace{y + \ldots + y}_{m}$$

as one can always take ∞ for y.

5.6. *2-ends compactification of $\mathbb{Z}$*

Consider the additive group $\mathbb{Z}$ with its natural embedding into the reals. A natural compactification of the real line adds two points, $+\infty$ and $-\infty$ with the obvious interpretation. It induces a compactification $\check{\mathbb{Z}}$ of $\mathbb{Z}$,

$$\check{\mathbb{Z}} = \mathbb{Z} \cup \{+\infty, -\infty\},$$

with the graph $S(x, y, z)$ of addition compactified so that

- $\models S(x, -\infty, -\infty)$, for all $x \neq +\infty$;

- $\models S(x, +\infty, +\infty)$, for all $x \neq -\infty$;
- $\models S(+\infty, -\infty, x)$, for any x.

The basic relations of language $\mathcal{C}$ are the relations defined from S by positive $\exists$-formulas.

Note that among the latter relations there are unary predicates, for all $n > 0$,

$$P_n(x) \equiv \exists y \, x = \underbrace{y + \ldots + y}_{n}.$$

Note that, for $n > 1$, $\neg P_n(q)$ holds.

Now we investigate for what ultrafilter D on $\mathbb{N}$ the family of finite cyclic groups $\mathbb{Z}/n\mathbb{Z}$, $n \in \mathbb{N}$, approximates $\check{\mathbb{Z}}$ along D. That is when

$$\lim_D \mathbb{Z}/n\mathbb{Z} = \check{\mathbb{Z}}. \tag{5.2}$$

Proposition 5.7: *(5.2) holds if and only if for any natural number m,*

$$\{n \in \mathbb{N} : m|n\} \in D. \tag{5.3}$$

Proof: Suppose the negation of (5.3) for some m, that is m does not divide n along the ultrafilter. Let $m = m_1 m_2$ such that $m_1|n$ and $(m_2, n) = 1$ for all $n \in X$, some $X \in D$, $m_2 \neq 1$. We may assume $m = m_2$. For all $n \in X$, let u_i, v_i be the integers such that $u_n m + v_n n = 1$. Correspondingly,

$$u_n m \equiv 1 \bmod n.$$

It follows

$$\prod_D \mathbb{Z}/n\mathbb{Z} \models \forall x P_m(x)$$

holds, in contrast with $\check{\mathbb{Z}} \models \neg P_m(1)$. So there is no homomorphism from the ultraproduct onto $\check{\mathbb{Z}}$.

Conversely, suppose (5.3) holds. Consider the ultrapower $^*\mathbb{Z} := \mathbb{Z}^N/D$ as an ordered additive group, $\mathbb{Z} \prec {^*}\mathbb{Z}$, $\mathbb{Z}$ is convex subgroup of $^*\mathbb{Z}$. Define, for $\eta \in {^*}\mathbb{Z}$,

$$\lim \eta = \begin{cases} +\infty, & \text{if } \eta > \mathbb{Z} \\ -\infty, & \text{if } \eta < \mathbb{Z} \\ m, & \text{if } \eta = m \in \mathbb{Z}. \end{cases}$$

This clearly is a homomorphism onto $\check{\mathbb{Z}}$ with respect to S and so all the relations in $\mathcal{C}$.

Note that D is a profinite ultrafilter by (5.3). By subsection 5.2, factoring by the torsion subgroup we get a surjective group homomorphism

$$\phi : \prod_D \mathbb{Z}/n\mathbb{Z} \to {}^*\mathbb{Z}.$$

Now we use the surjective homomorphism

$$\lim : {}^*\mathbb{Z} \to \check{\mathbb{Z}}$$

constructed above and finally the composition $\lim \circ \phi$ is a requred limit map.

□

6. Approximation by Some Finite Structures

6.1. *Approximation by finite fields*

In accordance with subsection 4.1 we discuss the approximation of a compactification $\bar{K} = K \cup \{\infty\} = \mathbf{P}^1(K)$, when speaking of an approximation of a field K. The standard topology that we will assume for $\bar{K}$ is the topology *generated by the Zariski topology on $\bar{K}$*, that is the smallest quasi-compact topology $\mathcal{T}$ extending the Zariski topology. Equivalently, by [22], these are the fields K such that for any elementary extension ${}^*K \succ K$ there is a specialisation (place) $\pi : {}^*\bar{K} \to \bar{K}$.

Remark 6.1: One may also consider the two-point compactification $\mathbb{R} \cup \{-\infty, +\infty\}$ of the field of reals. But if this is approximable, then so is $\bar{\mathbb{R}}$, since there exists an obvious surjective homomorphism $\mathbb{R} \cup \{-\infty, +\infty\} \to \bar{\mathbb{R}}$ taking $\pm\infty$ to ∞.

Conjecture: *For an infinite field, $\bar{K}$ is quasi-compact iff K is algebraically closed or K is isomorphic to one of the known non algebraically closed locally compact fields: $\mathbb{R}$ or finite extension of $\mathbb{Q}_p$ or $\mathrm{F}_p\{t\}$.*

Proposition 6.2: *(i) Any algebraically closed field K with respect to the Zariski language is approximable by finite fields.*

(ii) No locally compact field, other than algebraically closed, is approximable by finite fields.

Proof: (i) $\prod_D M_n = \mathrm{F}$, for M_n finite fields, is a pseudofinite field. Choose M_n and D so that $\mathrm{char}\,\mathrm{F} = \mathrm{char}\,K$. Let ${}^*\mathrm{F} \succ \mathrm{F}$ be a large enough elementary extension.

We will construct a total surjective specialisation $\pi : {}^*\mathrm{F} \to \bar{K} = K \cup \{\infty\}$. Obviously there is a partial specialisation, in fact embedding, of the prime field F_0 of charF into K. So we have constructed partial $\pi : {}^*F \to \bar{K}$. It is easy to extend π to the transcendence basis B of ${}^*F$ so that $\pi(B) = \bar{K}$, since algebraically independent elements satisfy no nontrivial Zariski closed relation.

Now note that any partial π into $\bar{K}$, for K algebraically closed, can be extended to a total one. This is the case when π is a partial specialisation from $\bar{K}'$ to $\bar{K}$ for K' algrbraically closed, since $\bar{K}$ s quasi-compact (see e.g. [15], Prop. 2.2.7) But ${}^*F \subseteq K'$ for some algebraically closed field, so the statement follows.

(ii) It is known ([24]) that F is a *pseudo-algebraically closed* field, that is any absolutely irreducible variety C over F has an F-point.

First we are going to prove (ii) for the case $K = \mathbb{R}$, the field of reals.

Claim: *The affine curve C given by the equations*

$$x^2 + y^2 + 1 = 0; \quad \frac{1}{x^2} + z^2 + 2 = 0$$

is irreducible over $\mathbb{C}$ and so is absolutely irreducible.

Proof: It is well known that $x^2 + y^2 + a = 0$, for $a \neq 0$, with any of the point removed is biregularly isomorphic to $\mathbb{C}$, and so irreducible. For the same reason the subvariety of $\mathbb{C}^2$ given by $\frac{1}{x^2} + z^2 + 2 = 0$ is also irreducible. We also note that the natural embeddings of both varieties into $\mathbf{P}^2$ are smooth.

The curve C projects into (x, y)-plane as the curve C_{xy} given by $x^2 + y^2 + 1 = 0$ and into the (x, z)-plane as the curve C_{xz} given by $\frac{1}{x^2} + z^2 + 2 = 0$.

Suppose towards a contradiction that $C = C_1 \cup C_2$ with C_1 an irreducible curve, $C_1 \neq C$, and C_2 Zariski closed. We denote $\bar{C}$, $\bar{C}_1$ and $\bar{C}_2$ the corresponding closures in the projective space $\mathbf{P}^3$.

Consider the projection $\mathrm{pr}_{xy} : \bar{C}_1 \to \bar{C}_{xy}$. This is surjective and the order of the projection is either 1 or 2. In the second case $\mathrm{pr}_{xy}^{-1}(a) \cap \bar{C}_1 = \mathrm{pr}_{xy}^{-1}(a) \cap \bar{C}$ for all $a \in \bar{C}_{xy}$, so $C = C_1$ and we are left with the first case only. In this case pr_{xy} is an isomorphism between $\bar{C}_1$ and $\bar{C}_{xy}$. It is also clear in this case that C_2 must be a curve, and pr_{xy} also an isomorphism from $\bar{C}_2$ to $\bar{C}_{xy}$. The points of intersection of C_1 and C_2 are the points over $a \in C_{xy}$ where $|\mathrm{pr}_{xy}^{-1}(a) \cap C| = 1$. One immediately sees that this can only be the points where $z = 0$, $x^2 = -\frac{1}{2}$, $y^2 = -\frac{1}{2}$.

We can apply the same arguments to the projection pr_{xz} onto $\bar{C}_{xz}$ and find that the points of intersection of C_1 and C_2 must satisfy $y = 0$, $x^2 = 1$ and $z^2 = -2$. The contradiction. Claim proved.

Now we prove that the existence of a total specialisation $\pi : {}^*F \to \mathbb{R} \cup \{\infty\}$ leads to a contradiction.

By above there exist a point (x, y, z) in $C(^*\mathrm{F})$. Then either $\pi(x)$ or $\pi(\frac{1}{x}) \in \mathbb{R}$ (are finite). Let us assume $\pi(x) \in \mathbb{R}$. Then necessarily $\pi(y) \neq \infty$, since $\pi(x)^2 + \pi(y)^2 + 1 = 0$, but the latter contradicts that $x^2 + y^2 \geq 0$ in $\mathbb{R}$. So (ii) for the reals is proved.

Now we prove (ii) for the remaining cases, that is locally compact nonarchimedean valued fields K. If L is a residue field for a valued field K, then the residue map $\bar{K} \to \bar{L}$ is a place. So assuming there is a sujective place $^*\mathrm{F} \to \bar{K}$ we get a surjective place $^*\mathrm{F} \to \bar{L}$. This is not possible for a PAC-field, by [24], Corollary 11.5.5. $\qquad\square$

6.2. *Approximation by finite groups*

In physics interesting gauge field theories are based on compact Lie groups such as the orthogonal group $\mathrm{SO}(3)$ or $\mathrm{SU}(N)$. On the other hand, since calculations in this theory and the analytic justification of the theory encounters enormous difficulties, there have been numerous attempts to develop a gauge field theory with finite group, see [25], or earlier [26] where an approximation of $\mathrm{SU}(3)$ by its finite subgroups was discussed.

The following, we believe, is crucial.

Problem 6.3: *Is the group* $\mathrm{SO}(3)$ *approximable by finite groups in the group language?*

More generally, let G be a compact simple Lie group. Is G approximable by finite groups in the group language? Equivalently (assuming for simplicity the continuum hypothesis), is there a sequence of finite groups G_n, $n \in \mathbb{N}$, an ultrafilter D on $\mathbb{N}$ and a surjective group homomorphism from the ultraproduct onto G,

$$\prod_D G_n \to G.$$

Remark 6.4: This problem has an easy solution (in fact, well-known to physicists) if we are content with G_n to be quasi-groups, that is omit the requirement of associativity of the group operation:

For each n, choose an $\frac{1}{n}$-dense finite subset $G(n) \subset G$ of points. For $a, b \in G(n)$ set $a * b$ to be a point in $G(n)$ which is at a distance less than $\frac{1}{n}$ from the actual product $a \cdot b$ in G. Now set, for $\gamma \in \prod_n G_n$,

$$\lim \gamma = g \text{ iff } \left\{ n \in \mathbb{N} : \ \mathrm{dist}(\gamma(n), g) \leq \frac{1}{n} \right\} \in D,$$

which is in fact the standard part map. Then clearly

$$\lim(\gamma_1 * \gamma_2) = \lim \gamma_1 \cdot \lim \gamma_2,$$

that is the map is a homomorphism.

Remark 6.5: By Theorem 4.5 for a given compact Lie group G the problem reduces to proving that for any positive sentence σ in the group language, such that $G \models \neg\sigma$, there is a finite group G_n such that $G_n \models \neg\sigma$.

Note that any compact simple Lie group G is definable in the field of reals in an explicit way, and hence the first order theory of G is decidable. This implies that the list of positive sentences σ such that $G \models \neg\sigma$, is recursive.

Finally, I would like to mention that in the last 2 years the problem was discussed with many people including M.Sapir, J.Wilson, C.Drutu and A.Muranov who made some valuable remarks, but no solution found as yet.

References

1. O.Alvarez, *Quantum mechanics and the index theorem* **Geometry and quantum field theory**, K.Uhlenbeck, D.Freed, editors Providence, R.I : American Mathematical Society, 1995, pp.185-270
2. Dominic Joyce, *Lectures on Calabi–Yau and special Lagrangian geometry*, arXiv:math/0108088v3, 2002
3. C.Isham and A.Doering, *A Topos Foundation for Theories of Physics: I. Formal Languages for Physics,* Journal of Mathematical Physics,49 (2007), Issue: 5, 36 pages; [quant-ph/0703060]
4. M.Morley, *Categoricity in Power,* Transactions of the American Mathematical Society, Vol. 114, No. 2, (Feb., 1965), pp. 514-538
5. R.J. Solomonoff, *A formal theory of inductive inference. Part I,* Infor. & Control 7 (1964):1
6. P.C.W. Davies, *Why is the Physical World So Comprehensible?* CTNS Bulletin 12.2, Spring 1992
7. P.C.W. Davies, *Algorithmic compressibility, fundamental and phenomenological laws.* **The Laws of Nature.** pp.248-267, Berlin: Walter de Gruyter (1995), ed. Friedel Weinert
8. E.Hrushovski, *A new strongly minimal set,* Annals of Pure and Applied Logic 62 (1993) 147-166
9. B.Zilber, *The structure of models of uncountably categorical theories.* In: **Proc.Int.Congr.of Mathematicians, 1983, Warszawa**, PWN - North-Holland P.Co., Amsterdam - New York - Oxford,1984, v.1, 359-368.
10. B.Zilber, *Pseudo-exponentiation on algebraically closed fields of characteristic zero*, Annals of Pure and Applied Logic, Vol 132 (2004) 1, pp 67-95
11. S.Shelah, **Classification Theory**, Lecture Notes in Mathematics, v.1292, 1987

12. B.Zilber, *Model theory, geometry and arithmetic of the universal cover of a semi-abelian variety.* In **Model Theory and Applications,** pp.427-458, Quaderni di matematica, v.11, Napoli, 2005

13. E.Hrushovski and B.Zilber, *Zariski Geometries.* Journal of AMS, 9 (1996), 1–96

14. A.Pillay and G.Srour, *Closed Sets and Chain Conditions in Stable Theories,* J. Symbolic Logic Volume 49, Issue 4 (1984), 1350-1362

15. B. Zilber, **Zariski Geometries,** Cambridge University Press, 2010.

16. B.Zilber *Non-commutative Zariski geometries and their classical limit,* Confluentes Mathematici, v.2,issue 2 , 2010 pp. 265-291

17. B. ZIlber, *A Class of Quantum Zariski Geometries,* in Model Theory with Applications to Algebra and Analysis, I, Volume 349 of LMS Lecture Notes Series, Cambridge University Press, 2008.

18. B.Zilber and V.Solanki, *Quantum Harmonic Oscillator as a Zariski Geometry,* arXiv:0909.4415

19. V.Solanki, **Zariski Structures in Noncommutative Algebraic Geometry and Representation Theory,** DPhil Thesis, University of Oxford, 2011

20. M.Rieffel, *Matrix algebras converge to the sphere for quantum Gromov–Hausdorff distance.* Mem. Amer. Math. Soc. 168 (2004) no. 796, 67-91; math.OA/0108005

21. G. Jaernefelt, *On the possibility of a finite model describing the universe,* Astronomische Nachrichten, vol. 297, no. 3, 1976, 131-139

22. B. Weglorz, *Equationally compact algebras*I. Fund. Math. 59 1966 289–298

23. P.Eklof and E.Fisher, *The elementary theory of Abelian groups,* Annals Math. Log. 4, 115–171

24. M.D. Fried and M. Jarden **Field arithmetic,** Second edition, revised and enlarged by Moshe Jarden, Ergebnisse der Mathematik (3) 11, Springer, Heidelberg,2005.

25. D.Freed and F.Quinn, *Chern-Simons theory with finite gauge group,* Comm. Math. Phys. , 156, no 3 (1993), 435-472

26. A. Bovier, M. Luling and D. Wyler, *Finite Subgroups Of Su(3),* J.Math.Phys. 22:1543, 1981

AN OBJECTIVE JUSTIFICATION FOR ACTUAL INFINITY?

Stephen G. Simpson

Department of Mathematics
Pennsylvania State University
State College, PA 16802, USA
simpson@math.psu.edu
http://www.math.psu.edu/simpson/

This document is a record of my contribution to a panel discussion which took place on July 27, 2011 as part of the Infinity and Truth Workshop held at the Institute for Mathematical Sciences, National University of Singapore, July 25–29, 2011.

1. Introduction

In preparation for the panel discussion, Professor Woodin asked each panelist to formulate a yes/no question to be asked of a benevolent, omniscient mathematician. In addition, each panelist was asked to give reasons for his choice of a question.

Since I do not believe in omniscient mathematicians, I chose to interpret "omniscient mathematician" as "wise and thoughtful philosopher of mathematics." With this change, my yes/no question reads as follows:

Can there be an *objective* justification for the concept "actual infinity"?

Of course this question would be incomprehensible without some understanding of the key terms "objective" and "actual infinity." Therefore, I shall now explain my views on objectivity in mathematics, and on potential infinity versus actual infinity. After that, I shall point to some relevant results from contemporary foundational research, especially reverse mathematics.

2. Objectivity in Mathematics

Generally speaking, by *objectivity* I mean human understanding of *reality*,[a] "the real world out there," with an eye toward controlling and using aspects of reality for human purposes. I subscribe to Objectivism [3], a well-known modern philosophical system which emphasizes the central role of objectivity.

My views on objectivity in mathematics are explained in my paper [6], which is the text of an invited talk that I gave at a philosophy of mathematics conference at New York University in April 2009. Briefly, I believe that mathematicians ought to seek objective understanding of the mathematical aspects of reality. This makes it possible to apply mathematics, with varying degrees of success, for the betterment of human life on earth.

Among the highly successful application areas for mathematics are: classical physics, engineering (mechanical engineering, electrical engineering, etc.), modern physics (relativity, quantum theory, etc.), chemistry, microbiology, astronomy. Among the successful application areas are: biology, medicine, agriculture, meteorology. Among the application areas with moderate to low success are: economics, social sciences, psychology, finance.

In all of these application areas, it is crucially important that our mathematical models should be *objective*, i.e., correspond closely to the underlying reality. Otherwise, success will be severely impaired. It would be desirable to place *all* of mathematics on an objective foundation. Failing that, it would be desirable to place at least the applicable parts of mathematics on an objective foundation.

3. Potential Infinity versus Actual Infinity

The distinction between potential infinity and actual infinity goes back to Aristotle. A detailed, nuanced discussion can be found in Books M and N of the Metaphysics [1, 2], which constitute Aristotle's treatise on the philosophy of mathematics. Aristotle's position is that, while potential infinities have an objective existence in reality, actual infinities do not. This is in the context of a broader argument against Plato's theology.

In modern mathematics, the prime example of *potential infinity* is the natural number sequence 1, 2, 3, ..., which manifests itself in reality as iteration, repeated processes, infinite divisibility,[b] etc. Another example in

---

[a]Of course reality is not limited to physical reality. For example, the United States government is a real entity but not a physical entity.

[b]For example, a piece of metal can be divided indefinitely.

modern mathematics is the full binary tree $\{0,1\}^{<\infty}$, whose infinite paths correspond roughly to the points on the unit interval $[0, 1]$.

Contrasted to potential infinity is *actual infinity*, i.e., a completed infinite totality. There are many examples in modern mathematics, including infinite sets such as $\omega = \{0, 1, 2, \ldots\}$, transfinite ordinals, $[0, 1]$, the real line, L_1, $B(H)$, etc. Thus my yes/no question comes down to asking whether certain parts of modern mathematics can have an objective justification.

4. Insights from Reverse Mathematics

As regards the distinction between potential and actual infinity, it appears that reverse mathematics can teach us something. Recall from [5] that *reverse mathematics* is a systematic attempt to classify specific mathematical theorms according to which set existence axioms are needed to prove them. The focus here is mainly on *core mathematics*, i.e., analysis, algebra, number theory, differential equations, probability, geometry, combinatorics, etc. Among the specific core mathematical theorems considered are many which time and again have proved useful in applications.

Reverse mathematics has uncovered a hierarchy of formal systems which are relevant for this classification. Some of the most important formal systems for reverse mathematics are, in order of increasing strength:

$$\mathsf{RCA}_0^*, \ \mathsf{RCA}_0, \ \mathsf{WKL}_0, \ \mathsf{ACA}_0, \ \mathsf{ATR}_0, \ \Pi_1^1\text{-}\mathsf{CA}_0, \ \ldots.$$

For our purposes here, recall [5] that there is a significant "break point" between the first three systems and the others. Namely, while RCA_0^*, RCA_0, WKL_0 are conservative over primitive recursive arithmetic ($= \mathsf{PRA}$) for Π_2^0 sentences, the other systems ACA_0, $\ldots$ are much stronger and therefore not conservative over PRA even for Π_1^0 sentences. Moreover, Tait [7] has argued that PRA represents the outer limits of finitism. Recall also that PRA is based on the idea of iteration and so may be viewed as a formal theory of potential infinity.

Now, an important discovery of reverse mathematics is that large parts of contemporary mathematics are formalizable in RCA_0^*, RCA_0, and WKL_0. This seems to include the applicable parts of mathematics. See also my paper [4]. Combining all of the above considerations, we see the possible outline of an objective justification of much of modern mathematics, especially the applicable parts of it. However, the prospects for an objective justification of actual infinity remain much more doubtful. This is the background of my yes/no question.

References

1. Hippocrates G. Apostle. *Aristotle's Philosophy of Mathematics.* University of Chicago Press, 1952. X + 228 pages.
2. Richard McKeon, editor. *The Basic Works of Aristotle.* Random House, 1941. XXXIX + 1487 pages.
3. Leonard Peikoff. *Objectivism: The Philosophy of Ayn Rand.* Dutton, New York, 1991. XV + 493 pages.
4. Stephen G. Simpson. Partial realizations of Hilbert's program. *Journal of Symbolic Logic,* 53:349–363, 1988.
5. Stephen G. Simpson. *Subsystems of Second Order Arithmetic.* Perspectives in Mathematical Logic. Springer-Verlag, 1999. XIV + 445 pages.
6. Stephen G. Simpson. Toward objectivity in mathematics. In this volume, pages 157–169.
7. William W. Tait. Finitism. *Journal of Philosophy,* 78:524–546, 1981.

ORACLE QUESTIONS

Theodore Slaman* and W. Hugh Woodin[†]

Department of Mathematics
University of California, Berkeley
Berkeley, CA 94720, USA
* slaman@math.berkeley.edu
† woodin@math.berkeley.edu

1. Introduction

The Infinity Workshop was focused on basic foundational questions. The participants were given the following lead questions prior to the meeting.

- What is the nature of mathematical truth and how does one resolve questions which are formally unsolvable within ZFC, such as the Continuum Hypothesis?
- Do the discoveries in Mathematics provide evidence favoring one philosophical view, such as Platonism or Formalism, over others?

During the meeting, we held an unusual problem session. Rather than asking the participants to share open problems to drive mathematical progress, we asked them to share questions to drive foundational progress and to facilitate this discussion we asked following question, inviting the workshop participants to respond.

You have an audience with the Oracle of Mathematics. You can ask one Yes-No question. What is your question?

Here we have the results. Some of these are exactly as given in the Oracle session; some have been altered or elaborated by the participants upon reflection.

2. Questions

2.1. *Ilijas Farah*

Question 2.1: Is there a physical experiment whose outcome is independent from *ZFC*?

2.2. *Moti Gitik*

Question 2.2: $Con(\exists$ strongly compact$) \iff Con(\exists$ supercompact$)$?

2.3. *Joel David Hamkins*

Question 2.3: Are we correct in thinking we have an absolute concept of the finite?

I might mischievously ask the question my six-year-old daughter Hypatia often puts to our visitors: *"Answer yes or no. Will you answer 'no'?"* They stammer, caught in the liar paradox, as she giggles. But my actual question is:

Are we correct in thinking we have an absolute concept of the finite?

An absolute concept of the finite underlies many mathematician's understanding of the nature of mathematical truth. Most mathematicians, for example, believe that we have an absolute concept of the finite, which determines the natural numbers as a unique mathematical structure—$0, 1, 2$, and so on—in which arithmetic assertions have definitive truth values. We can prove after all that the second-order Peano axioms characterize $\langle \mathbb{N}, S, 0 \rangle$ as the unique inductive structure, determined up to isomorphism by the fact that 0 is not a successor, the successor function S is one-to-one and every set containing 0 and closed under S is the whole of $\mathbb{N}$. And to be finite means simply to be equinumerous with a proper initial segment of this structure. Doesn't this categoricity proof therefore settle the matter?

I don't think so. The categoricity proof, which takes place in set theory, seems to my way of thinking merely to push the absoluteness question for finiteness off to the absoluteness question for sets instead. And surely this is a murkier realm, where already mathematicians do not universally agree that we have a single absolute background concept of set. We know by forcing and other means how to construct alternative set concepts, which seem fully as legitimate and set-theoretic as the set concepts from which they are derived. Thus, we have a plurality of set concepts, and our confidence in a unique absolute set-theoretic background is weakened. How then can we sensibly base our confidence in an absolute concept of the finite on set theory? Perhaps this absoluteness is altogether illusory.

My worries are put to rest if the oracle should answer positively. A negative answer, meanwhile, would raise alarms. A negative answer could indicate, on the one hand, that our understanding of the finite is simply incoherent, a catastrophe, where our cherished mathematical theories are all inconsistent. But, more likely in my view, a negative answer could also mean that there is an undiscovered plurality of concepts of the finite. I imagine technical developments arising that would provide us with tools to modify the arithmetic of a model of set theory, for example, with the same power and flexibility that forcing currently allows us to modify higher-order features, while not providing us with any reason to prefer one arithmetic to another (unlike our current methods with non-standard models). The discovery of such tools would be an amazing development in mathematics and lead to radical changes in our conception of mathematical truth.

2.4. *Juliette Kennedy*

Question 2.4: Is there a natural/absolute/transcendent notion of proof?

In his 1946 Princeton Bicentennial Lecture Gödel suggested the problem of finding notions of provability and of definability for set theory, which are not "dependent on the formalism chosen." What Gödel is actually suggesting there is duplicating the Turing analysis of the notion of computable function — a notion which is very robust with respect to its various associated formalisms — in the cases of provability and definability. One way to interpret this suggestion vis a vis definability is to consider standard notions of definability in set theory, which are usually built over first order logic, and change the underlying logic. It turns out that constructibility is not very sensitive to the underlying logic, and the same goes for hereditary ordinal definability; this means that a large class of logics can be substituted for first order logic in the construction of L, without changing L, and the same holds for second order logic in the case of HOD. (Joint work with Magidor and Väänänen.) The question I asked at the conference was whether there is a similar, formalism free notion of proof in set theory? Would this involve an analysis similar to the above, based on a change of logic? Or would a transfer of the Turing analysis to the case of provability involve a different test of robustness? Gödel suggested in the lecture that

"It is not impossible that ... some completeness theorem would hold, which would say that every proposition expressible in set theory is decidable from the present axioms plus some true assertion about the largeness of the universe of all sets." We can regard Woodin's theorem on Σ_1^2 absoluteness as a result of this kind.

2.5. Steffen Lempp

Question 2.5: Are the computably enumerable Turing degrees an atomic model of their theory?

A positive answer would provide a partial solution a more widely circulated conjecture that each c.e. Turing degree is definable; so this question may be viewed as a step toward this widely believed conjecture.

2.6. Stephen G. Simpson

Question 2.6: Does there exist an *objective* justification for the concept of *actual infinity*?

2.7. Theodore Slaman

Question 2.7: Is there a non-trivial automorphism of the Turing degrees?

There are two fundamental open questions concerning the structural properties of relative computability. The first question is whether there is a non-trivial automorphism of the partial ordering of Turing degrees, D. If not, then D is bi-interpretable second-order arithmetic, thereby reducing the logical properties of the Turing degrees to those of the representatives of those degrees. The second question is to settle Martin's Conjectured classification of the functions from 2^ω to 2^ω which are Turing invariant, i.e. induce functions from D to D. Martin Conjectured that AD implies that the non-constant such functions are well-ordered under eventual point-wise domination with successor given by the Turing jump. If so, then the degree invariant functions give a very fine hierarchy for the structure of arbitrary relative definability. See [2] for more information.

2.8. Jouko Väänänen

Question 2.8: Is there a largest tree of size $\aleph_1$ without uncountable branches?

2.9. *Nik Weaver*

Question 2.9: Is there a consistent theory of QED in 3+1 dimensions?

A good background reference is [1].

2.10. *W. Hugh Woodin*

Question 2.10: It the following a true axiom for V?

(1) There is a supercompact cardinal.
(2) Suppose that ϕ is a Σ_3-sentence and ϕ is true in V. Then, there is a universally Baire set $A \subset \mathbb{R}$ such that $(HOD)^{L(A,\mathbb{R})} \cap V_\Theta \models \phi$, where Θ is $\Theta^{L(A,\mathbb{R})}$.

A *weak extender model* for δ is a supercompact is simply a transitive inner model N of ZFC containing all the ordinals such that δ is supercompact in N and this is witnessed by the supercompactness of δ in V in the following sense. For each $\gamma > \delta$, there is a normal fine δ-complete ultrafilter U on $\mathcal{P}_\delta(\gamma)$ such that

$$\mathcal{P}_\delta(\gamma) \cap N \in U$$

and such that $U \cap N \in N$. By the results of [3], N is necessarily *close* to V above δ and moreover N inherits essentially all large cardinals which hold in V above δ (at least up to the level of Axiom I0). The *suitable extender models* of [3] are simply a refinement of weak extender models designed to pass through the I0 barrier, [5].

My question for the ORACLE is motivated by the following conjecture where

$$\text{``}V = \text{Ultimate-}L\text{''}$$

is the axiom defined below.

Definition ($V = $ Ultimate-L).

(1) There is a strong cardinal and a proper class of Woodin cardinals.
(2) For each Σ_3-sentence ϕ, if ϕ holds in V then there is a universally Baire set $A \subseteq \mathbb{R}$ such that

$$\text{HOD}^{L(A,\mathbb{R})} \cap V_\Theta \models \phi$$

where $\Theta = \Theta^{L(A,\mathbb{R})}$. $\qquad\qquad \square$

Conjecture (Ultimate-L Conjecture). *Suppose there is an extendible cardinal. Then there exists a suitable extender model $\mathbb{M}$ such that*

(1) $\mathbb{M} \subseteq \mathrm{HOD}$,
(2) $\mathbb{M} \models$ "$V = \text{Ultimate-}L$".   □

If the Ultimate-L Conjecture is true then one has a compelling candidate for an axiom for V. Hence the question.

There are some known consequences of the axiom, "$V = \text{Ultimate-}L$". The Generic-Multiverse is the generic-multiverse generated by V, [4].

Theorem ($V = \text{Ultimate-}L$).

(1) CH *holds.*
(2) *The Ω Conjecture holds.*
(3) V *is the minimum universe of the Generic-Multiverse.*   □

2.11. *Boris Zilber*

Question 2.11: Is there a pseudo-finite group G that has a homomorphism onto $SU(3)$?

Same question with $SU(3)$ replaced by any simple compact Lie group. This question is about an approximation by finite groups of gauge groups, which is of fundamental importance in quantum field theory.

References

1. Rudolf Haag. *Local quantum physics*. Texts and Monographs in Physics. Springer-Verlag, Berlin, second edition, 1996. Fields, particles, algebras.
2. Andrew Marks, Theodore A. Slaman, and John R. Steel. Martin's conjecture, arithmetic equivalence, and countable Borel equivalence relations. preprint, 2011.
3. W. Hugh Woodin. Suitable Extender Models I. *Journal of Mathematical Logic*, 10(1-2):101–341, 2010.
4. W. Hugh Woodin. The Continuum Hypothesis, the generic-multiverse of sets, and the Ω Conjecture. In *Set Theory, Arithmetic and Foundations of Mathematics: Theorems, Philosophies*, volume 36 of *Lecture Notes in Logic*, pages 13–42. Cambridge University Press, New York, NY, 2011.
5. W. Hugh Woodin. Suitable Extender Models II: beyond ω-huge. *Journal of Mathematical Logic*, 11(2):115–437, 2011.